Operational Amplifier Noise

Operational Amplifier Noise

*Techniques and Tips for Analyzing
and Reducing Noise*

Art Kay

ELSEVIER

AMSTERDOM • BOSTON • HEIDELBERG • LONDON
NEW YORK • OXFORD • PARIS • SAN DIEGO
SAN FRANCISCO • SINGAPORE • SYDNEY • TOKYO

Newnes is an imprint of Elsevier

Newnes

Newnes is an imprint of Elsevier
225 Wyman Street, Waltham, MA 02451, USA
The Boulevard, Langford Lane, Kidlington, Oxford OX5 1GB, UK

Notice
No responsibility is assumed by the publisher for any injury and/or damage to persons
or property as a matter of products liability, negligence or otherwise, or from any use
or operation of any methods, products, instructions or ideas contained in the material
herein. Because of rapid advances in the medical sciences, in particular, independent
verification of diagnoses and drug dosages should be made

Library of Congress Cataloging-in-Publication Data
A catalog record for this book is available from the Library of Congress

British Library Cataloguing-in-Publication Data
A catalog record for this book is available from the British Library

ISBN: 978-0-7506-8525-2

For information on all Newnes publications
visit our website at www.elsevierdirect.com

Typeset by MPS Limited, a Macmillan Company, Chennai, India
www.macmillansolutions.com

Contents

Preface

Noise can be defined as any unwanted signal in an electronic system. Noise is responsible for reducing the quality of audio signals or introducing errors into precision measurements. Board- and system-level electrical design engineers are interested in determining the worst case noise they can expect in their design, design methods for reducing noise, and measurement techniques to accurately verify their design.

Intrinsic and extrinsic noise are the two fundamental types of noise that affect electrical circuits. Extrinsic noise is generated by external sources. Digital switching, 60 Hz noise, and power supply switching are common examples of extrinsic noise. Intrinsic noise is generated by the circuit element itself. Broadband noise, thermal noise, and flicker noise are the most common examples of intrinsic noise. This book describes how to predict the level of intrinsic noise in a circuit with calculations, and using Spice simulations. Noise measurement techniques are also discussed.

Acknowledgments

Special thanks for all of the technical insights from the following individuals:

Texas Instruments

Rod Burt, Senior Analog IC Design Manager
Bruce Trump, Manager – Linear Products
Matt Hann, Applications Engineering Manager
Bryan Zhao, Field Applications Engineering
Neil Albaugh, Senior Applications Engineer Retired

Analog and Rf Models

Bill Sands, Consultant

Cirrus Logic

Tim Green, Senior Staff Systems Engineer

www.en-genius.net

Paul McGoldrick, Editor-in-Chief

CHAPTER 1

Introduction and Review of Statistics

Noise analysis can be done in the time domain, in the frequency domain, or by using statistical analysis. This chapter introduces these three analysis methods. These methods will be utilized throughout this book.

1.1 Time Domain View of Noise

Noise is most commonly viewed in the time domain. A typical plot of time domain noise is shown in Figure 1.1. In the time domain, noise voltage is on the *y*-axis and time is on the *x*-axis. Noise can be viewed in the time domain using an oscilloscope. Figure 1.1 also shows that if you look at this random signal statistically, it can be represented as a Gaussian distribution. The distribution is drawn sideways to help show its relationship with the time domain signal. In Section 1.2, the statistical view is discussed in detail.

Figure 1.1 shows thermal noise in the time domain. Thermal noise is generated by the random motion of electrons in a conductor. Because this motion increases with temperature, the magnitude of thermal noise increases with temperature. Thermal noise can be viewed as a random variation in the voltage present across a component (e.g., resistor). Figure 1.2 gives the equation for finding the root mean square (RMS) thermal noise given resistance temperature and bandwidth.

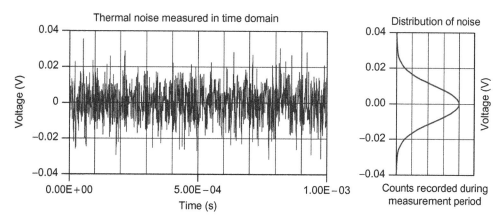

Figure 1.1: Thermal noise shown in the time domain and statistically

$$e_n = \sqrt{4 \cdot k \cdot T \cdot R \cdot \Delta f} \quad \text{(1.1) Thermal noise equation}$$

where
e_n is the RMS noise voltage
T is temperature in Kelvin (K)
R is resistance in Ohms (Ω)
Δf is noise bandwidth in Hertz (Hz)
k is Boltzmann's constant (J/K)

Note: To convert degrees Celsius to Kelvin

$$T_k = 273.15\ °C + T_c$$

Figure 1.2: Thermal noise voltage equation

$$f(x) = \frac{1}{\sigma \cdot \sqrt{2\pi}} \cdot e^{[-(x-\mu)^2/2\sigma^2]} \quad \begin{array}{l}\text{(1.2) Probability density function} \\ \text{for normal (Gaussian) distribution}\end{array}$$

where
f(x) is the probability that x will be measured at any instant in time
x is the random variable
μ is the mean value
σ is the standard deviation

Figure 1.3: Probability density function for a Gaussian distribution

The important thing to know about the thermal noise equation is that it allows you to find an RMS noise voltage. In many cases, engineers want to know the peak-to-peak noise voltage. In Section 1.2, we will learn some statistical methods that can be used to estimate peak-to-peak noise voltage given the RMS noise value. Section 1.2 also covers other basic statistical methods that are used in noise analysis.

1.2 Statistical View of Noise

Most forms of intrinsic noise have a Gaussian distribution and can be analyzed using statistical methods. For example, statistical methods must be used to calculate the sum of two noise signals or estimate peak-to-peak amplitude. This section gives a short review of some basic statistical methods required to carry out noise analysis.

1.2.1 Probability Density Function

The mathematical equation that forms the normal distribution function is called the "probability density function" (see Figure 1.3). Plotting a histogram of noise voltage measured over a time interval will approximate the shape of this function. Figure 1.4 illustrates a measured noise histogram with the probability distribution function superimposed on it.

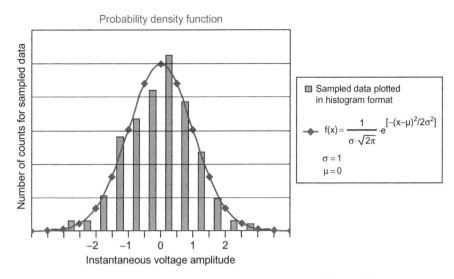

Figure 1.4: Measured distribution with superimposed probability density function

$$P(a<x<b)=\int_{a}^{b} f(x)\,dx=\int_{a}^{b} \frac{1}{\sigma\cdot\sqrt{2\pi}}\cdot e^{[-(x-\mu)^2/2\sigma^2]} dx$$

(1.3) Probability distribution function for normal (Gaussian) distribution

where

$P(a<x<b)$ is the probability that x will be in the interval (a, b) in any instant in time. For example, $P(-1<x<+1)=0.3$ means that there is a 30% chance that x will be between −1 and 1 for any measurement

x is the random variable. In this case noise voltage

μ is the mean value

σ is the standard deviation

Figure 1.5: Probability distribution function

1.2.2 Probability Distribution Function

The probability distribution function is the integral of the probability density function. This function is very useful because it tells us about the probability of an event that will occur in a given interval (see Figures 1.5 and 1.6. For example, assume that Figure 1.6 is a noise probability distribution function. The function tells us that there is a 30% chance that you will measure a noise voltage between $-1\,V$ and $+1\,V$ [i.e., the interval $(-1, 1)$] at any instant in time.

This probability distribution function is instrumental in helping us translate RMS to peak-to-peak voltage or current noise. Note that the tails of the Gaussian distribution are infinite. This implies that any noise voltage is possible. While this is theoretically true, in practical

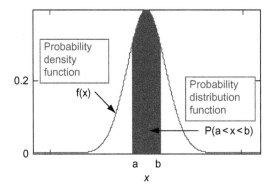

Figure 1.6: Probability density function and probability distribution function

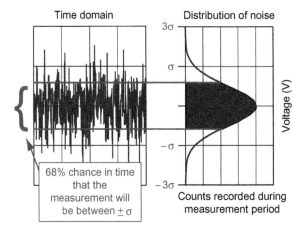

Figure 1.7: Illustrates how standard deviation relates to peak-to-peak

terms the probability that extremely large instantaneous noise voltages are generated is very small. For example, the probability that we measure a noise voltage between -3σ and $+3\sigma$ is 99.7%. In other words, there is only a 0.3% chance of measuring a voltage outside of this interval. For this reason, $\pm3\sigma$ (i.e., 6σ) is often used to estimate the peak-to-peak value for a noise signal. Note that some engineers use 6.6σ to estimate the peak-to-peak value of noise. There is no agreed-upon standard for this estimation. Figure 1.7 graphically shows how 2σ will catch 68% of the noise. Table 1.1 summarizes the relationship between standard deviation and probability of measuring a noise voltage.

Thus, we have a relationship that allows us to estimate peak-to-peak noise given the standard deviation. In general, however, we want to convert RMS to peak-to-peak amplitude. Often, people assume that the RMS and standard deviation are the same. This is not always the case. The two values are equal only when there is no DC component (the DC component is

Table 1.1: Number of Standard Deviations and Chance
of Measuring Voltage

Number of Standard Deviations	Chance of Measuring Voltage (%)
2σ (same as $\pm\sigma$)	68.3
3σ (same as $\pm 1.5\sigma$)	86.6
4σ (same as $\pm 2\sigma$)	95.4
5σ (same as $\pm 2.5\sigma$)	98.8
6σ (same as $\pm 3\sigma$)	99.7
6.6σ (same as $\pm 3.3\sigma$)	99.9

the average value μ). In the case of thermal noise, there is no DC component, so the standard deviation and RMS values are equal.

One way of computing the RMS noise voltage is to measure a large number of discrete points and use statistics to estimate the standard deviation. For example, if you have a large number of samples from an analog-to-digital (A/D) converter, you could use Eqs. (1.4)–(1.6) to compute the mean, standard deviation and RMS of the noise signal. Many software packages such as Microsoft Excel can be used to compute these functions. In fact, Excel has built-in functions for computing standard deviation and mean value. Some test equipment will include these and other built-in mathematical functions. For example, many oscilloscopes include RMS, mean and standard deviation functions.

In general, it is best to use the standard deviation function as apposed to RMS when doing noise computations. Intrinsic noise should not have a DC component. In some practical cases, the instrument measuring the noise may have a DC component. The DC component should not be included in the noise computation because it is not really part of the noise signal. When using the RMS formula on a noise signal with a DC component, the results will be affected by the DC component. The standard deviation formula, however, will eliminate the effects of the DC component.

One final statistical concept to cover is the addition of noise signals. To add two noise signals, we must know if the signals are correlated or uncorrelated. Noise signals from two independent sources are uncorrelated. For example, the noise from two independent resistors or two operational amplifiers (op-amps) is uncorrelated. A noise source can become correlated through a feedback mechanism. Noise-canceling headphones are a good example of the addition of correlated noise sources. They cancel acoustic noise by summing inversely correlated noise. Eq. (1.4) shows how to add correlated noise signals. Note that in the case of the noise-canceling headphones, the correlation factor would be $C = -1$. (see Figure 1.8).

In most cases, we will add uncorrelated noise sources (see Eq. (1.5)). Adding noise in this form will effectively sum two vectors using the Pythagorean theorem. Figure 1.9 shows the addition graphically. A useful approximation is that if one of these sources is one-third the amplitude of the other, the smaller source can be ignored.

$$e_{nT} = \sqrt{e_{n1}^2 + e_{n2}^2 + 2C \cdot e_{n1} \cdot e_{n2}}$$

(1.4) Addition of two correlated noise sources

$$e_{nT} = \sqrt{e_{n1}^2 + e_{n2}^2}$$

(1.5) Addition of two uncorrelated noise sources

where

C is a correlation factor $-1 < C < +1$

e_{n1} is a noise source

e_{n2} is a noise source

Figure 1.8: Addition of random uncorrelated signals

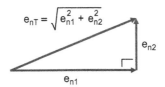

$$e_{nT} = \sqrt{e_{n1}^2 + e_{n2}^2}$$

e_{n2}

e_{n1}

Figure 1.9: Pythagorean theorem for noise

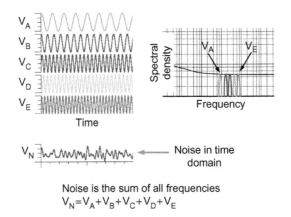

Noise is the sum of all frequencies
$$V_N = V_A + V_B + V_C + V_D + V_E$$

Figure 1.10: Understanding the spectral density curve

1.3 Frequency Domain View of Noise

An important characteristic of noise is its spectral density. Voltage noise spectral density is a measurement of RMS noise voltage per square root hertz (or commonly nV/$\sqrt{\text{Hz}}$). Power spectral density is given in W/Hz. A random noise signal can be thought of as an infinite summation of sine waves at different frequencies. The right side of Figure 1.10 shows how several signals at different frequency add to form a random noise signal. The left side of Figure 1.10 shows the same signals in the frequency domain. Note that each sine wave

$$\mu = \frac{1}{n} \sum_{i=1}^{n} x_i \qquad \text{(1.6) Mean value}$$

$$\sigma = \sqrt{\sigma^2} = \sqrt{\frac{1}{n} \sum_{i=1}^{n} (x_i - \mu)^2} \qquad \text{(1.7) Standard deviation}$$

$$RMS = \sqrt{\frac{1}{n} \sum_{i=1}^{n} x_i^2} \qquad \text{(1.8) RMS}$$

Figure 1.11: Statistical equations for a discrete population

$$e_n = \sqrt{4k \cdot T \cdot R \cdot \Delta f}$$

$$\frac{e_n}{\sqrt{\Delta f}} = \sqrt{4k \cdot T \cdot R} \qquad \text{(1.9) Thermal noise equation}$$

Figure 1.12: Thermal noise equation rearranged into spectral density format

creates an impulse or "spike" in the frequency domain. The example shows just five sine waves combining to form a "random" signal. In reality, noise signals have infinite frequency components. You can imagine an infinite number of impulses in the frequency domain combine to form the spectral density curve.

Earlier, we have learned that the thermal noise of a resistor can be computed using Eq. (1.1). This equation can be rearranged into a spectral density form (see Figure 1.11). One important characteristic of this noise is that it has a flat spectral density plot (i.e., it has uniform energy at all frequencies). For this reason, thermal noise is sometimes called broadband noise or white noise (Figure 1.12). The word "white" is used to describe noise with a uniform energy at all frequencies because white light is generated by mixing all colors (wavelengths) with uniform energy. Op-amps also have broadband noise associated with them. Broadband noise is defined as noise that has a flat spectral density plot. Figure 1.13 shows the noise spectral density of a resistor graphed vs. resistance. This plot can be used as a quick way to determine the spectral density of a resistance. Also note that temperature has a very small effect on overall noise.

Figure 1.14 shows two common regions in spectral density curves. In the broadband region the noise spectral density is flat, so the contribution of all the different frequency components is equal. Op-amps also may have a low-frequency noise region that does not have a flat spectral density plot. This noise is called 1/f noise, flicker noise, or low-frequency noise. Typically, the power spectrum of 1/f noise falls at a rate of 1/f. This means that the voltage

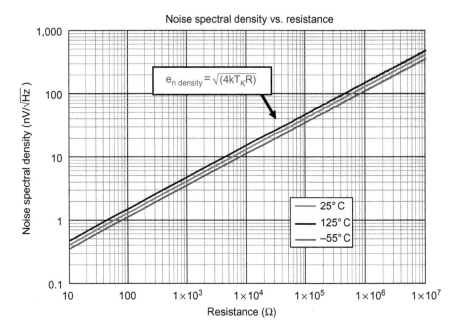

Figure 1.13: Thermal noise equation rearranged into spectral density format

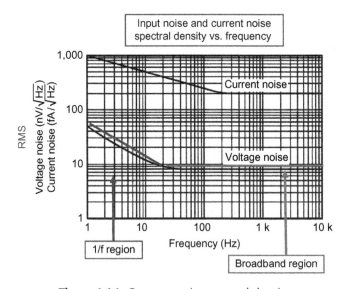

Figure 1.14: Op-amp noise spectral density

spectrum falls at a rate of $1/f^{(1/2)}$. In practice, however, the exponent of the 1/f function may deviate slightly. Figure 1.14 shows a typical op-amp spectrum with both a 1/f region and a broadband region. Note that the spectral density plot also shows current noise (given in $fA/\sqrt{Hz}$).

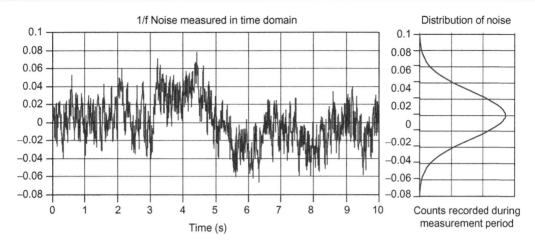

Figure 1.15: 1/f Noise shown in the time domain and statistically

$$\text{Noise power} = \int_{f_1}^{f_2} e_n^2 \, df$$

(1.10) RMS noise power over a frequency range.
e_n is the noise voltage spectral density.

$$\text{Noise voltage} = \sqrt{\int_{f_1}^{f_2} e_n^2 \, df}$$

(1.11) RMS noise voltage over a frequency range.
e_n is the noise voltage spectral density.

Figure 1.16: Converting spectral density to RMS noise

Note that 1/f noise also has a normal distribution, and consequently the mathematics described earlier still applies. Figure 1.15 shows the time domain description of 1/f noise. Note that the *x*-axis of this graph is given in seconds; this slow change with time is typical for 1/f noise.

1.4 Converting Spectral Density to RMS Noise

A very common noise calculation is to convert spectral density to RMS noise. This is used extensively in op-amp noise calculations. There are three different types of spectral densities to consider: noise power spectral density (unit W/Hz), noise voltage spectral density (unit V/$\sqrt{\text{Hz}}$), and noise current spectral density (unit V/$\sqrt{\text{Hz}}$). Noise power spectral density can be converted to RMS power by integrating the spectral density. Power spectral density is defined as voltage or current spectral density squared. Thus, to convert voltage or current noise spectral density to RMS noise, you convert to power (v_n^2 or i_n^2), integrate, and convert back to voltage or current (square root). See Figure 1.16 for details.

A common error that people make when converting voltage spectral density to RMS noise voltage is integrating the voltage spectral density instead of the power spectral density.

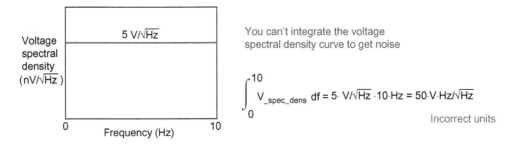

Figure 1.17: Incorrect way to compute noise

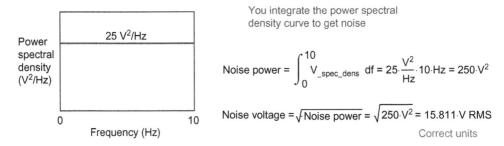

Figure 1.18: Correct way to compute noise

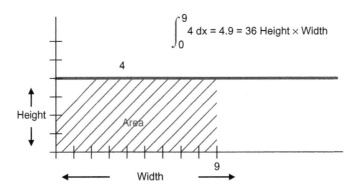

Figure 1.19: Integration computes area under a curve

In Figures 1.17 and 1.18, we will do a dimensional analysis to demonstrate why this does not work. Before doing this, however, we will review the integral. As a quick reminder, the integral function will give the area under a curve. Figure 1.19 shows how a constant function can be integrated by simply multiplying the height by the width (i.e., the area of a rectangle). Considering the integral to be the area of a rectangle simplifies the conversion of spectral density curves to RMS noise values.

Figure 1.17 shows the strange units that result when you attempt to integrate the voltage spectral density curve. Figure 1.18 shows how you can integrate the power spectral density and convert back to voltage by taking the square root of the result. Note that we get the proper units. Also note that power spectral density is simply the voltage or current spectral density squared (remember $P = V^2/R$ and $P = I^2R$). Thus, by looking at Figures 1.17 and 1.18, you can see why power spectral density must be integrated rather than voltage or current spectral density.

Chapter Summary

- Oscilloscope measurements show noise in the time domain.
  - x-axis is time and y-axis is voltage or current.
- A Gaussian distribution is a statistical view of noise.
  - Standard deviation and RMS are the same if noise does not have a DC component (average value is zero).
  - Six times the standard deviation is a good estimate of peak-to-peak noise.
  - There is a 99.7% probability that noise will be less than six times the standard deviation.
  - Most noise is uncorrelated.
  - Uncorrelated noise is added with the root sum of the square of each noise component.
- The noise spectral density is a frequency domain view of noise.
  - Spectral density has units of $V/\sqrt{Hz}$ or $A/\sqrt{Hz}$.
  - White noise is composed of an infinite number of different frequency components with equal energy.
  - Two key regions in a noise spectral density curve are the 1/f region and the broadband region.
  - Noise spectral density can be converted to RMS noise by taking the square root of the integral of the noise signal squared.

Questions

1.1 Calculate the RMS thermal noise for a 10-kΩ resistor with a 10-kHz noise bandwidth.

1.2 Estimate the peak-to-peak noise for a 10-kΩ resistor with a 10-kHz noise bandwidth.

1.3 Calculate the spectral density for a 10-kΩ resistor.

1.4 A 10 mV-RMS and a 5 mV-RMS noise source are in series with each other. Assuming the noise sources are random, what is the total noise?

1.5 A 10-mV RMS and a 5-mV RMS noise source are in series with each other. Assuming the noise sources are correlated and the correlation factor is -1, what is the total noise?

1.6 A 25-mVpp noise signal is measured. What is the approximate RMS voltage?

Further Reading

Hogg, R.V., Tanis, E.A., 1988, Probability and Statistical Inference, third ed. Macmillan, London.
Motchenbacher, C.D., Connelly, J.A., 1993, Low-Noise Electronic System Design, Wiley-Interscience, New York.

CHAPTER 2

Introduction to Op-Amp Noise

2.1 Op-Amp Noise Analysis Technique

The goal of op-amp noise analysis technique is to calculate the peak-to-peak output noise of an op-amp circuit based on op-amp data sheet information. As the technique is explained, we will use formulas that apply to most simple op-amp circuits. For more complex circuits, the formulas can help to get a rough idea of the expected noise output. It is possible to develop more accurate formulas for these complex circuits; however, the math would be overly complex. For the complex circuits, it is probably best to use a three-step approach. First, get a rough estimate using the formulas; second, get a more accurate estimate using Spice; and finally, verify your results through measurements.

As the example circuit, we will use a simple noninverting amplifier with a Texas Instruments OPA277 (see Figure 2.1). Our goal is to determine the peak-to-peak output noise. To do this, we have to consider the op-amp current noise, the op-amp voltage noise, and the resistor thermal noise. We will determine the value of these noise sources using the spectral density curves in the data sheet. Also, we will have to consider the gain and bandwidth of the circuit.

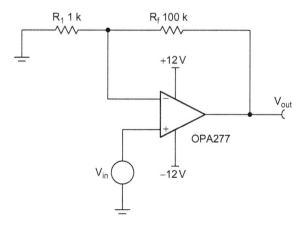

Figure 2.1: Example circuit for noise analysis

Operational Amplifier Noise.

13

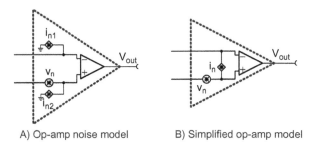

A) Op-amp noise model   B) Simplified op-amp model

Figure 2.2: Op-amp noise model

2.2 Introducing the Op-Amp Noise Model

Two different models for op-amp noise are shown in Figure 2.2. Figure 2.2A consists
of two uncorrelated current noise sources and one voltage noise source connected to the
op-amp's inputs. The voltage noise source can be thought of as time-varying input offset
voltage component, and the current noise sources can be thought of time-varying bias
current components. Note that in some cases, the magnitude of i_{n1} and i_{n2} can be different.
In Figure 2.2B, the two current noise sources are combined into a single noise source
common to both inputs. The model shown in Figure 2.2B is most commonly used for
noise calculations.

2.3 Noise Bandwidth

Integration of the power spectral density curve for the voltage and current spectra will give
us the RMS magnitude of the sources in the op-amp model (Figure 2.2). However, the shape
of the spectral density curve will contain a 1/f region and a broadband region with a low-pass
filter (see Figure 2.3). Calculation of the total noise of these two sections will require the use
of formulas that were derived using calculus. The results of these two computations are added
using root sum of square (RSS) addition for uncorrelated sources, which has been discussed
in Chapter 1.

First, we will integrate the broadband region with a low-pass filter. Ideally, the low-pass filter
portion of this curve would be a straight vertical line. This is referred to as a brick wall filter.
Solving the area under a brick wall filter is easy because it is a rectangle (height × width). In
the real world, we cannot realize a brick wall filter. However, there are a set of constants that
can be used to convert real-world filter bandwidth to an equivalent brick wall filter bandwidth
for the purpose of the noise calculation. Figure 2.4 compares the theoretical brick wall filter
to first-, second-, and third-order filters.

Figure 2.5 shows the equations to convert the real-world filter or the brick wall equivalent.
Table 2.1 lists the brick wall conversion factors (K_n) for different filter orders. For example, a

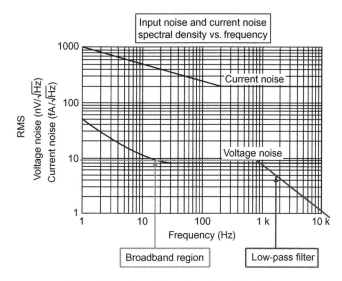

Figure 2.3: Broadband region with filter

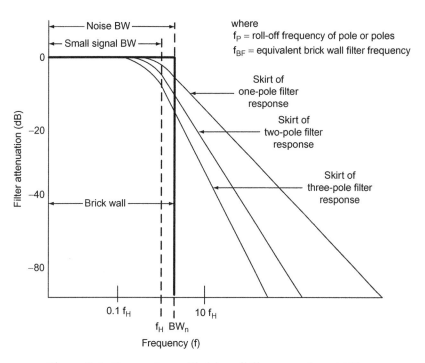

Figure 2.4: Comparison of brick wall filter to real-world filter

$$BW_n = f_H \cdot K_n \quad (2.1) \quad \text{Effective noise bandwidth}$$

where

f_H is the upper-cut frequency
K_n is the brick wall conversion factor

Figure 2.5: Noise bandwidth for simple filter on broadband region

Table 2.1:

Number of Poles in Filter	Kn AC Noise Bandwidth Ratio
1	1.57
2	1.22
3	1.16
4	1.13
5	1.12

$$E_{nBB} = e_{BB} \cdot \sqrt{BW_n}$$ (2.2) Broadband noise
voltage in volts RMS

e_{BB} Broadband noise voltage density (V/$\sqrt{Hz}$)

BW_n Noise bandwidth for a given system (Hz)

Figure 2.6: Broadband noise equation

first-order filter bandwidth can be converted to a brick wall filter bandwidth by multiplying by 1.57. The adjusted bandwidth is sometimes referred to as the noise bandwidth. Note that the conversion factor approaches one as the order increases. In other words, higher order filters are a better approximation of a brick wall filter.

Now that we have a formula to convert a real-world filter to its brick wall equivalent, it is a simple matter to integrate the power spectrum. Remember, integrating the power is the voltage spectrum squared. At the end of the integration, the square root is taken to convert back to voltage.

2.4 Broadband RMS Noise Calculation

Eq. (2.2) shown in Figure 2.6 gives the relationship for converting broadband noise spectral density to RMS noise. The broadband noise spectral density (e_{BB}) is taken from the op-amp data sheet. The noise bandwidth is derived using Eq. (2.1) from Table 2.1 (Figures 2.7 and 2.8).

2.5 1/f RMS Noise Calculation

Recall that our goal is to determine the magnitude of the noise source V_n from Figure 2.3. This noise source consists of both broadband noise and 1/f noise. Using Eqs. (2.2) and (2.3), we have been able to compute the broadband component. Now we need to compute the 1/f component. This is done by integrating the power spectrum of the 1/f region of the noise spectral density plot. Figure 2.9 shows this region graphically. The result of the integration is given by Eqs. (2.4) and (2.5). Eq. (2.4) normalizes any noise measurement in the 1/f region to the noise at 1 Hz. In some cases, this number can be read directly from the chart; in other

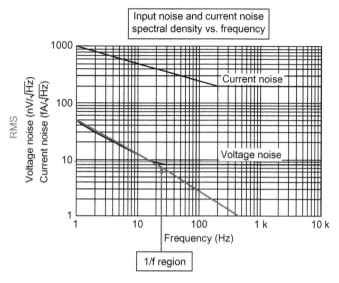

Figure 2.7: 1/f Region

$$e_{fnorm} = e_{at\_f} \cdot \sqrt{f} \qquad (2.3) \text{ 1/f noise normalized to 1 Hz}$$

e_{at_f} is voltage noise density selected at a frequency in the 1/f region

f is a frequency in the 1/f region where the noise density is known

Figure 2.8: Noise at 1 Hz (normalized)

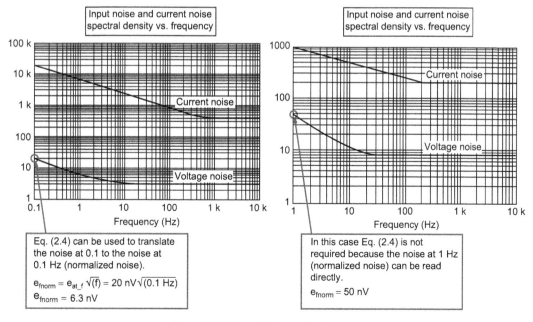

Figure 2.9: Two 1/f normalizing cases

$$E_{nf} = e_{fnorm} \cdot \sqrt{\ln\left(\frac{f_H}{f_L}\right)}$$ (2.4) 1/f noise measured in volts RMS

where

e_{fnurm} is normalized noise at 1 Hz from Eq. (2.3)

f_H is upper frequency of operation (use the noise bandwidth BW_n)

f_L is lower frequency of operation (0.1 Hz is typically used)

Figure 2.10: 1/f Noise computation

cases, it is more convenient to use this equation (see Figure 2.10). Eq. (2.5) computes the 1/f noise using the normalized noise, upper noise bandwidth, and lower noise bandwidth. The full derivation is given in Section 2.12.

When considering the 1/f noise you must choose a low-frequency cutoff. This is because the 1/f function is not defined at zero (i.e., 1/0 is undefined). In fact, the noise theoretically goes to infinity when you integrate back to zero hertz. However, you should consider that very low frequencies correspond to long times. For example, 0.1 Hz corresponds to 10 s, and 0.001 Hz corresponds to 1000 s. For extremely low frequencies, the corresponding time could be years (e.g., 10 nHz = 3 years). The greater the frequency interval that is integrated, the larger the resultant noise. Keep in mind, however, that extremely low frequency noise measurements must be made over a long period of time. These phenomena will be discussed in greater detail in Chapter 8. For now, please note that 0.1 Hz is often used for the lower cutoff frequency of the 1/f calculation.

2.6 Combining Flicker and Broadband Noise

Now we have both the broadband and the 1/f noise magnitude. We must add these noise sources using the formula for uncorrelated noise sources given in Chapter 1 (see Eq. (2.5) and Eq. (1.8)).

A common concern that engineers have when considering this analysis technique is that they feel that the 1/f noise and broadband noise should be integrated in two separate regions. In other words, they believe that adding noise in this region will create an error because the 1/f noise will add with the broadband noise outside of the "1/f region." In reality the 1/f region extends across all frequencies as does the broadband region. Note that the noise spectrum is shown on a log chart, so the 1/f region has little impact after it drops below the broadband curve. The only region where the combination of the two curves is obvious is near where they combine (often called the 1/f corner frequency). In this region, we can see that the two sections combine as is described by our mathematical model. Figure 2.12 illustrates how the two regions actually overlap as well as giving some relative magnitudes.

$$E_{n\_v} = \sqrt{E_{nf}^2 + E_{nBB}^2}$$ (2.5) Total RMS voltage noise

where
E_{nf} is total RMS 1/f noise
E_{nBB} is total RMS broadband noise

Figure 2.11: Addition of 1/f and broadband noise

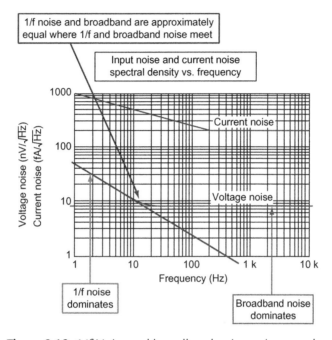

Figure 2.12: 1/f Noise and broadband noise regions overlap

2.7 Noise Model for Example Circuit

In Chapter 1, we have developed a method for converting the noise spectral density curves from a product data sheet to noise sources in an op-amp model. In this chapter, we will learn how to use the model to compute the total output noise for a simple op-amp circuit. The total noise referred-to-the-input (RTI) will contain noise from the op-amp voltage noise source, noise from the op-amp current noise source, and resistor noise. This combined noise source will be multiplied by the op-amp "noise gain." Figure 2.13 shows all the different sources needing to be combined and multiplied by the noise gain.

2.8 Noise Gain

Noise gain is the gain that the op-amp circuit has to the total noise RTI. In some cases this is not equivalent to the signal gain. Figure 2.14 shows an example where the signal gain is one and the noise gain is two. The V_n source represents contributions of noise from several sources; it is positioned on the noninverting input ("+ve" input) of the op-amp circuit. Note

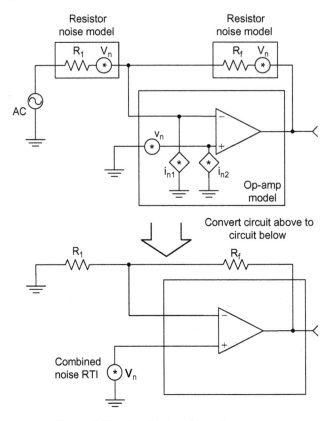

Figure 2.13: Combining the noise sources

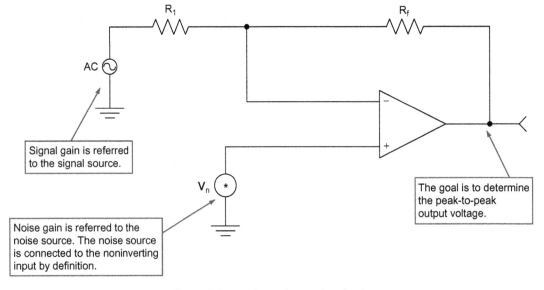

Figure 2.14: Noise gain vs. signal gain

$$\text{Noise gain} = \frac{R_f}{R_1} + 1$$

(2.6) Noise gain is the gain seen by the noise voltage source

Figure 2.15: Noise gain for simple op-amp circuit

$$I_{nBB} = i_{nBB} \cdot \sqrt{BW_n}$$

(2.7) Broadband RMS noise current

$$R_{eq} = \frac{R_f \cdot R_1}{R_f + R_1}$$

(2.8) Equivalent resistance seen by current noise source

$$i_{fnorm} = i_{at\_f} \cdot \sqrt{f}$$

(2.9) 1/f noise normalized to 1 Hz

$$I_{nf} = i_{fnorm} \cdot \sqrt{\ln\left(\frac{f_H}{f_L}\right)}$$

(2.10) 1/f noise measured in amps RMS

$$I_n = \sqrt{I_{nf}^2 + I_{nBB}^2}$$

(2.11) Total RMS noise from current noise

$$E_{n\_i} = I_n \cdot R_{eq}$$

(2.12) RMS noise voltage from current noise

where

i_{nBB} is current noise broadband spectral density

i_{at_f} is current noise spectral density in 1/f region at f

f is frequency that the current spectral density i_{at_f} is measured at

BW_n is noise bandwidth (2.1)

R_f is feedback resistor

R_1 is input resistor

Figure 2.16: Converting current noise to voltage noise (RTI) for simple op-amp

that it is a common engineering practice to lump all the noise sources to a common source at the noninverting input. Our end goal is to compute noise referred-to-the-output (RTO) of the op-amp circuit (Figure 2.15).

2.9 Converting Current Noise to Voltage Noise

From the previous chapter we know how to compute the voltage noise input, but how do we convert the current noise sources to a voltage noise source? One way of doing this is to carry out an independent nodal analysis for each current source and use superposition to sum the results. Be careful to make sure that the results from each current source is added using RSS. Eqs. (2.7)–(2.12) in Figure 2.16 allow you to convert current noise to an equivalent voltage noise source for a simple op-amp circuit. Figure 2.17 shows the noise equivalent circuit.

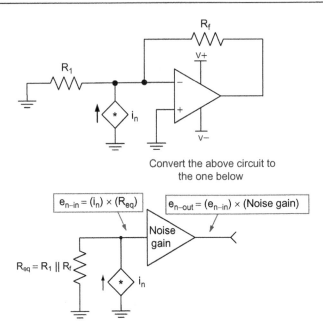

Figure 2.17: Converting current noise to voltage noise (equivalent circuit)

$$E_{n\_r} = \sqrt{4kT \cdot R_{eq} \cdot BW_n}$$ (2.13) Total RMS noise from thermal
noise of op-amp feedback network

$$R_{eq} = \frac{R_f \cdot R_1}{R_f + R_1}$$ (2.8) Equivalent resistance seen
by current noise source

where
k is Boltzmann's constant (1.38×10^{-23} J/K)
T is temperature in Kelvin (K)
R_{eq} is equivalent resistance
BW_n is noise bandwidth

Figure 2.18: Thermal noise RTI for simple op-amp circuit

2.10 Including the Effect of Thermal Noise

Another thing that must be considered is the thermal voltage noise from the resistors
in the op-amp circuit. Figure 2.18 gives the equations for thermal noise. From an
AC perspective, R_f and R_1 are in parallel, so the equivalent resistance given in Eq.
(2.8) can be used to find the thermal noise equivalent resistance. Eq. (2.13) uses the
equivalent resistance and noise bandwidth to compute thermal noise. This input referred

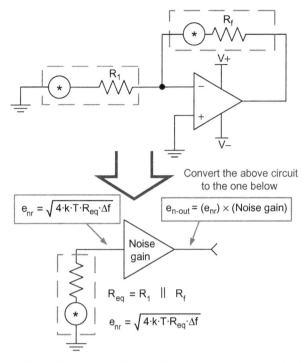

Figure 2.19: Thermal noise RTI for simple op-amp circuit (equivalent circuit)

$$E_{n\_in} = \sqrt{E_{n\_i}^2 + E_{n\_v}^2 + E_{n\_r}^2}$$ (2.14) Total RMS noise referred to the input

$$E_{n\_out} = E_{n\_in} \cdot \text{Noise\_gain}$$ (2.15) Total RMS noise referred to the output

$$E_{n\_out\_pp} = E_{n\_out} \cdot 6.0$$ (2.16) Peak-to-peak output voltage

Figure 2.20: Conversion to peak-to-peak noise

thermal noise source is expressed as an equivalent resistor. Figure 2.19 shows the noise equivalent circuit.

2.11 Combining All the Noise Sources and Computing Peak-to-Peak Output Noise

The final step to compute output noise is to combine all the noise sources and multiply by the noise gain. This RMS noise is used to estimate the peak-to-peak noise by multiplying by 6. Recall from Chapter 1 that there is a 99.7% chance that any instantaneous noise measurement will be less than six times the RMS noise. Eqs. (2.14)–(2.16) from Figure 2.20 summarize this final step.

2.12 Derivation of Key Noise Formulas

Figure 2.21 shows the derivation of the brick wall correction factor used when computing noise bandwidth. This derivation involves integrating the power spectral density and taking the square root of the result.

$$E_{RMS}^2 = \int_{f_1}^{f_2} e_n^2 \cdot (|G|)^2 \, df$$

where

E_{RMS} is total RMS noise from f_1 to f_2 in V_{RMS}
e_n is magnitude of noise spectral density at f_1 in $V/\sqrt{Hz}$

G is gain function for a single pole filter

$$G = \frac{1}{1 + \frac{j\omega}{\omega_p}} \qquad |G| = \frac{1}{\sqrt{1^2 + \frac{f^2}{f_p^2}}} \qquad (|G|)^2 = \frac{1}{1^2 + \frac{f^2}{f_p^2}}$$

$$E_{RMS}^2 = \int_{f_1}^{f_2} e_n^2 \cdot \frac{1}{1 + \frac{f^2}{f_p^2}} \, df = \int_{f_1}^{f_2} e_n^2 \cdot \frac{f_p^2}{f_p^2 + f^2} \, df$$

$$E_{RMS}^2 = e_n^2 \cdot f_p \cdot atan\left(\frac{f_2}{f_p}\right) - e_n^2 \cdot f_p \cdot atan\left(\frac{f_1}{f_p}\right)$$

Let $f_1 = 0$, $f_2 = \infty$

$$E_{RMS}^2 = e_n^2 \cdot f_p \cdot atan(\infty) - e_n^2 \cdot f_p \cdot atan\left(\frac{f_1}{f_p}\right) = e_n^2 \cdot f_p \cdot \frac{\pi}{2}$$

$$E_{RMS}^2 = e_n^2 \cdot f_p \cdot \frac{\pi}{2}$$

$$E_{RMS} = \sqrt{e_n^2 \cdot f_p \cdot \frac{\pi}{2}}$$

Note that π/2 is K_n = 1.57 from Table 2.1

Figure 2.21: Derivation of brick wall correction factor

$$E_n = \frac{e_{normal}}{\sqrt{f}}$$

$$E_n^2 = \frac{e_{normal}^2}{\left(\sqrt{f}\right)^2} = \frac{e_{normal}^2}{f}$$

$$E_{nf}^2 = \int_{f_L}^{f_H} \frac{e_{normal}^2}{f}\, df = e_{normal}^2 \cdot \left(\ln(f_H) - \ln(f_L)\right)$$

$$E_{nf}^2 = e_{normal}^2 \cdot \ln\left(\frac{f_H}{f_L}\right)$$

$$E_{nf} = e_{normal} \cdot \sqrt{\ln\left(\frac{f_H}{f_L}\right)} \qquad \text{(2.4) Equation for calculating RMS noise in 1/f region}$$

Figure 2.22: Derivation of 1/f region

$$\text{Output\_noise}^2 = \sqrt{e_{n1}^2 + e_{n2}^2} = \sqrt{\left(e_{n1} \cdot \frac{R_f}{R_1}\right)^2 + e_{n2}^2}$$

Let $\quad \beta = 4 \cdot kT \cdot BW$

$$\text{Output\_noise} = \sqrt{\left(\sqrt{\beta \cdot R_1} \cdot \frac{R_f}{R_1}\right)^2 + \left(\sqrt{\beta \cdot R_f}\right)^2}$$

$$\text{Input\_noise} = \frac{\sqrt{\left(\sqrt{\beta \cdot R_1} \cdot \frac{R_f}{R_1}\right)^2 + \left(\sqrt{\beta \cdot R_f}\right)^2}}{\frac{R_f}{R_1} + 1}$$

$$\text{Input\_noise}^2 = \frac{\beta \cdot \frac{R_f^2}{R_1} + \beta \cdot R_f}{\frac{\left(R_f + R_1\right)^2}{R_1^2}} = \frac{\beta \cdot R_f^2 \cdot R_1 + \beta \cdot R_f \cdot R_1^2}{\left(R_f + R_1\right)^2}$$

$$\text{Input\_noise} = \sqrt{\frac{\beta \cdot R_f^2 \cdot R_1 + \beta \cdot R_f \cdot R_1^2}{\left(R_f + R_1\right)^2}} = \sqrt{\beta \cdot \frac{R_f \cdot R_1}{R_f + R_1}}$$

$$\text{Input\_noise} = \sqrt{4 \cdot kT \cdot BW \cdot \left(\frac{R_f \cdot R_1}{R_f + R_1}\right)} \qquad \text{Eq. to noise of } R_f \parallel R_1$$

Figure 2.23: The equivalent noise resistance

Figure 2.22 shows the derivation for the 1/f noise region. This derivation involves integrating the power spectral density and taking the square root of the result.

This calculation shows the combination of the two thermal noise sources from each resistor in an op-amp circuit. The two sources are combined using superposition and the final result is algebraically rearranged to prove that the equivalent resistance is the parallel combination of R_f and R_1. This makes sense because the two resistors are in parallel from an AC perspective and noise is an AC signal (Figure 2.23).

Chapter Summary

- The simplified op-amp noise model contains current noise and voltage noise.
- Noise bandwidth of a simple filter can be calculated by multiplying the signal bandwidth (i.e., -3-dB bandwidth) by a brick wall correction factor.
- Noise calculations involve looking at the effects of 1/f noise and broadband noise separately. The two components are added using the root sum of squares.
- The final noise calculation takes into account current noise, voltage noise, and thermal noise.

Questions

2.1   Given a signal bandwidth of 1 kHz, what is the noise bandwidth? Assume a first-order filter.

2.2   Given the spectral density curve below:
  a.   What is the total noise contribution of the 1/f region? Assume $f_L = 0.1$ Hz and $f_H = 1$ kHz.
  b.   What is the total noise contribution of the broadband region? Assume the signal bandwidth is 1 kHz and the filter is first order.

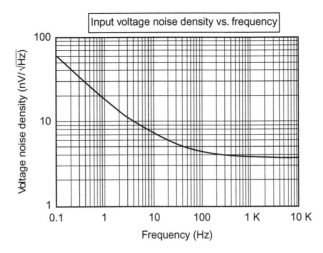

2.3　For the circuit below:

    a.　What is the thermal noise? Assume $BW_n = 5\,kHz$.

    b.　What is the current noise referred-to-the-input. Assume in $= 5\,fA/\sqrt{Hz}$, and $BW_n = 5\,kHz$.

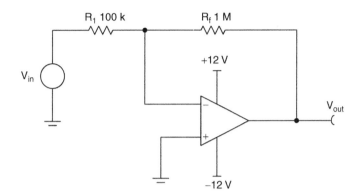

2.4　Assume the total noise at the output of an op-amp contains $E_{n-i} = 100\,\mu V_{RMS}$, $E_{n-v} = 150\,\mu V_{RMS}$, and $E_{n-r} = 75\,\mu V_{RMS}$. What is the total noise?

Further Reading

Robert V. Hogg, and Elliot A, 1988. Tanis. Probability and Statistical Inference, third ed. Macmillan, London.

C. D. Motchenbacher, and J. A. Connelly, 1993. Low-Noise Electronic System Design. Wiley-Interscience, New York.

CHAPTER 3

Op-Amp Noise Example Calculations

This chapter covers two detailed noise hand calculations. These calculations use the equations developed in Chapter 2. The concept of the gain bandwidth limiting the op-amp frequency response is introduced. Also, a multiple-stage amplifier example shows how the input stage will be the dominant noise source.

3.1 Example Calculation #1: OPA627 Noninverting Amplifier

At this point, we are finally ready to go through a real-world example. Sometimes engineers are overwhelmed by the amount of work required to get to this point. In fact, it is possible to use simulation software to do some of this difficult work for you. However, it is important to have an understanding of the theoretical background because it will give you a more intuitive understanding of how noise works. Furthermore, you should always do a quick "back-of-the-envelope" calculation before you simulate a circuit so that you know if your simulation result is correct. In Chapter 4, we will discuss how to do this analysis using a Spice simulator package.

Figure 3.1 illustrates the simple op-amp configuration that will be used for this analysis example. Note that the specifications used in this example were taken from the OPA627 data sheet. This data sheet can be downloaded from the Texas Instruments website (http://www.ti.com).

3.2 Compute the Noise Bandwidth

The first step to this analysis is to determine the noise gain and noise bandwidth for this circuit. The noise gain is given by Eq. (3.2) (Noise_Gain $=R_f/R_1+1=100k/1k+1=101$). The signal bandwidth is limited by the closed loop bandwidth of the op-amp. Using the gain bandwidth product from the data sheet, the closed loop bandwidth can be determined using Eq. (3.1). If the gain bandwidth product is not listed in the data sheet, use the unity-gain bandwidth specification. The unity-gain bandwidth is the same as gain bandwidth for unity-gain stable amplifiers. Figure 3.2 shows the equations used to calculate bandwidth. Figure 3.5 shows graphically how bandwidth is limited by Aol.

3.3 Get Key Noise Specifications from the Data Sheet

The next part of the analysis is to get the broadband and 1/f noise spectral density specification from the data sheet. The specification is sometimes shown graphically

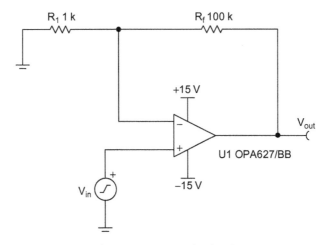

Figure 3.1: Example circuit

$$\text{Closed\_Loop\_Bandwidth} = \frac{\text{Gain\_Bandwidth\_Product}}{\text{Noise\_Gain}} \qquad \text{(3.1) Op-amp bandwidth limit}$$

$$\text{Closed\_Loop\_Bandwidth} = \frac{16 \text{ MHz}}{101} = 158 \text{ kHz}$$

Figure 3.2: Closed loop bandwidth for simple noninverting amp

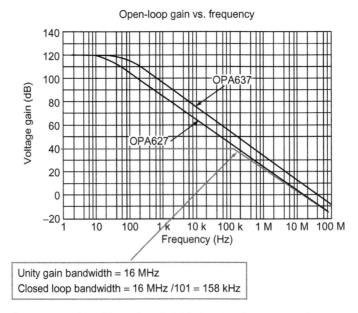

Figure 3.3: Closed loop bandwidth for simple noninverting amp

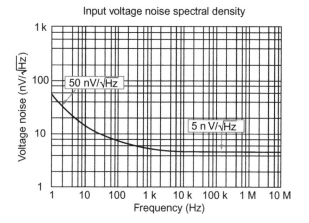

Figure 3.4: OPA627 noise spectral density specification to be used in calculations

Computer total op-amp voltage noise contribution:

Broadband voltage noise component

$$BW_n = f_H \cdot K_n \qquad\qquad \text{(2.1) Noise bandwidth}$$

$$BW_n = (158 \text{ kHz}) \cdot (1.57) = 248 \text{ kHz}$$

$$E_{nBB} = e_{BB} \cdot \sqrt{BW_n} \qquad\qquad \text{(2.2) Total broadband noise from voltage noise}$$

$$E_{nBB} = \left(5\frac{nV}{\sqrt{Hz}}\right) \cdot \sqrt{248 \text{ kHz}} = 2490 \text{ nV}_{RMS}$$

1/f Voltage noise component:

$$e_{fnorm} = e_{at\_f} \cdot \sqrt{f} \qquad\qquad \text{(2.3) 1/f Noise normalized to 1 Hz}$$

$$e_{fnorm} = \left(50\frac{nV}{\sqrt{Hz}}\right) \cdot \sqrt{1 \text{Hz}} = 50 \text{ nV}$$

$$E_{nf} = e_{fnorm} \cdot \sqrt{\ln\left(\frac{f_H}{f_L}\right)} \qquad\qquad \text{(2.4) Total 1/f noise from voltage noise}$$

$$E_{nf} = (50 \text{ nV}) \cdot \sqrt{\ln\left(\frac{248 \text{ kHz}}{0.1 \text{ Hz}}\right)} = 191.8 \text{ nV}_{RMS}$$

Total voltage noise (referred to the input of the amplifier):

$$E_{n\_v} = \sqrt{E_{nf}^2 + E_{nBB}^2} \qquad\qquad \text{(2.5) Total RMS noise from voltage noise}$$

$$E_{n\_v} = \sqrt{(2490 \text{ nV}_{RMS})^2 + (191.8 \text{ nV}_{RMS})^2} = 2497 \text{ nV}_{RMS}$$

Figure 3.5: Computation of the magnitude of the voltage noise referred-to-the-input

Table 3.1: OPA627 Noise Spectral Density Specification

| | OPA627AM, AP, AU OPA637AM, AP, AU | | | |
	Min	Typ	Max	Units
Input voltage noise				
Noise density: f = 10 Hz		20		nV/√Hz
f = 100 Hz		10		nV/√Hz
f = 1 kHz		5.6		nV/√Hz
f = 10 kHz		4.8		nV/√Hz
Voltage noise, BW = 0.1–10 Hz		0.8		µVpp
Input bias current noise				
Noise density, f = 100 Hz		2.5		fA/√Hz
Current noise, BW = 0.1–10 Hz		48		fApp

Broadband current noise component

$$BW_n = f_H \cdot K_n \tag{2.1}$$

$$BW_n = (158\ kHz) \cdot (1.57) = 248\ kHz$$

$$I_{nBB} = i_{nBB} \sqrt{BW_n} \tag{2.7}$$

$$I_{nBB} = \left(2.5 \frac{fA}{\sqrt{Hz}}\right) \cdot \sqrt{248\ kHz} = 1.244\ pA_{RMS}$$

$$I_{nf} = 0\ A \qquad \text{No 1/f noise}$$

$$I_n = \sqrt{I_{nf}^2 + I_{nBB}^2} \tag{2.11}$$

$$I_n = 1.244\ pA_{RMS}$$

$$R_{eq} = R_1 \parallel R_f$$

$$R_{eq} = \frac{R_f \cdot R_1}{R_f + R_1} = \frac{(100\ k\Omega) \cdot (1\ k\Omega)}{100\ k\Omega + 1\ k\Omega}$$

$$E_{n\_i} = I_n \cdot R_{eq} \tag{2.12}$$

$$E_{n\_i} = (1.244\ pA_{RMS})(0.99\ k\Omega) = 1.23\ nV_{RMS}$$

Figure 3.6: Conversion of current noise spectral density to equivalent input noise voltage

(see Figure 3.6) or in a table format (see Figure 3.7). The spectral density values and the closed loop bandwidth are used to compute the total input voltage noise. Example 3.1 shows how the total input noise is computed using the formulas introduced previously.

Thermal noise (resistor noise) component

$$BW_n = f_H \cdot K_n \tag{2.1}$$

$$BW_n = (158 \text{ kHz}) \cdot (1.57) = 248 \text{ kHz}$$

$$R_{eq} = R_1 \parallel R_f$$

$$R_{eq} = \frac{R_f \cdot R_1}{R_f + R_1} = \frac{(100 \text{ k}\Omega) \cdot (1 \text{ k}\Omega)}{100 \text{ k}\Omega + 1 \text{ k}\Omega} \tag{2.8}$$

$$E_{n\_r} = \sqrt{4 \cdot k \cdot T \cdot R_{eq} \cdot \Delta f} \tag{2.13}$$

$$E_{n\_r} = \sqrt{4 \left(1.38 \cdot 10^{-23} \text{ J/K}\right) (298 \text{ K}) \cdot (0.99 \text{ k}\Omega) \cdot (248 \text{ kHz})} = 2010 \text{ nV}_{RMS}$$

Since the total voltage noise is $E_{n\_v} = 2497$ nV$_{RMS}$, the resistor noise voltage (2010 nV$_{RMS}$) is significant.

Figure 3.7: Conversion of resistor noise to equivalent input noise voltage

3.4 Compute Total Op-Amp Voltage Noise Contribution

3.4.1 Compute Total Op-Amp Current Noise Contribution

Now we need to convert the current noise to an equivalent referred-to-the-input (RTI) voltage noise. First, we will convert the current noise spectral density to a current source. This current source is multiplied by an equivalent input resistance to compute input voltage noise. It should be noted that the 1/f calculation is not required for this example because the amplifier is a JFET input. JFET amplifiers generally do not have 1/f current noise. This procedure is summarized in Figure 3.6.

3.5 Compute Total Thermal Noise Contribution

Figure 3.7 illustrates how RTI resistor noise is calculated. Note that for this example, the resistor noise is similar in magnitude to the op-amp noise and so it will significantly contribute to the output noise.

3.6 Combine All the Noise Sources and Compute Peak-to-Peak Output

Now that we have computed all the noise components, we can determine the total noise RTI. This result will be multiplied by the noise gain to compute the noise referred to the output.

Voltage noise form op-amp RTI:

$E_{n\_v} = 2497\ nV_{RMS}$ Figure 3.5

Current noise form op-amp RTI:

$E_{n\_i} = 1.23\ nV_{RMS}$ Figure 3.6

Resistor noise form op-amp RTI:

$E_{n\_r} = 2010\ nV_{RMS}$ Figure 3.7

Total RMS noise RTI:

$$E_{n\_in} = \sqrt{E_{n\_v}^2 + E_{n\_i}^2 + E_{n\_r}^2} \qquad (2.14)$$

$$E_{n\_in} = \sqrt{(2497\ nV_{RMS})^2 + (1.23\ nV_{RMS})^2 + (2010\ nV_{RMS})^2} = 3205\ nV_{RMS}$$

Total RMS noise RTO:

$$E_{n\_out} = E_{n\_in} \cdot Noise\_Gain \qquad (2.15)$$

$$E_{n\_out} = (3205\ nV_{RMS})\ (101) = 324\ \mu V_{RMS}$$

Estimate total peak-to-peak noise RTO:

$$E_{n\_out\_pp} = E_{n\_out} \cdot 6 \qquad (2.16)$$

$$E_{n\_out\_pp} = (324\ \mu V_{RMS})\cdot 6 = 1.94\ mV_{pp}$$

Figure 3.8: Computation of total peak-to-peak output noise

Finally, the conversion factor from Table 1.1 will be used to estimate the peak-to-peak output. Figure 3.8 shows the details.

3.7 Example Calculation #2: Two-Stage Amplifier

Example noise calculation #2 will analyze the noise for a two-stage amplifier. The goal of this example is to show how the equations derived in Chapter 2 can be applied to a different circuit topology. This circuit has source impedance R_1 associated with V_{in}. The circuit also has two low-pass filters (R_4C_1 and R_7C_2) (Figure 3.9).

Figure 3.10 shows the calculations for the total noise from the input amplifier's (U1) noise voltage source. This noise contains both 1/f and broadband noise. This noise does not include any effects of the second-stage amplifier (U2). The noise bandwidth in this example is limited by the two filters. Because the cut frequency (f_c) of the two filters is the same, they act as a single second-order filter ($K_n = 1.22$).

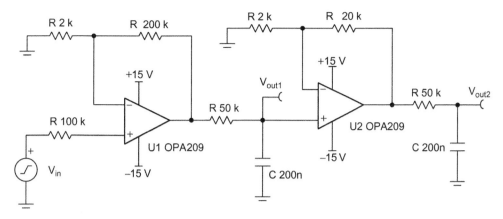

Figure 3.9: Two-stage amplifier for noise example

Gain_Bandwidth_Product := 18 MHz   From data sheet

Noise_Gain: = 101

$$\text{Closed\_Loop\_Bandwidth} := \frac{\text{Gain\_Bandwidth\_Product}}{\text{Noise\_Gain}} = 178.2 \times 10^3 \text{ Hz}$$

$$f_c := \frac{1}{2\pi \cdot (50 \text{ k}\Omega) \cdot (200 \text{ nF})} = 15.9 \text{ Hz}$$

f_c < Closed_Loop_Bandwidth so f_c is used for noise bandwidth

$$BW_n := 1.22 \, f_c = 19.4 \text{ Hz}$$

$$e_n := 2.2 \frac{\text{nV}}{\sqrt{\text{Hz}}} \quad \text{From data sheet}$$

$$E_{nBB} := e_n \cdot \sqrt{BW_n} = 9.69 \times 10^{-9} \text{ V} \quad \text{Total broadband noise}$$

$$f_L := 0.1 \text{ Hz} \quad e_{at\_f} := 20 \frac{\text{nV}}{\sqrt{\text{Hz}}} \quad \text{From data sheet}$$

$$e_{fnorm} := e_{at\_f} \cdot \sqrt{f_L} = 6.32 \times 10^{-9} \text{ V}$$

$$E_{nf} := e_{fnorm} \cdot \sqrt{\ln\left(\frac{BW_n}{f_L}\right)} = 14.52 \times 10^{-9} \text{ V} \quad \text{Total 1/f noise}$$

$$E_{n\_v} := \sqrt{E_{nBB}^2 + E_{nf}^2} = 17.46 \times 10^{-9} \text{ V} \quad \text{Total noise from voltage noise}$$

Figure 3.10: Total noise from first-stage op-amp noise voltage

Figure 3.11 shows the calculations for the total noise from the input amplifier's (U1) noise current source. The current noise flows through both the feedback network and the source impedance R_1. This noise does not include any effects of the second-stage amplifier (U2).

Figure 3.12 shows the calculations for the total noise from the input amplifier's (U1) resistor noise. Both the feedback network (R_2 and R_3) and the source impedance (R_1) contribute to the noise. This noise does not include any effects of the second-stage amplifier (U2).

Figure 3.13 shows the combination of the current voltage and resistor noise for the input stage (U1). In this example, the current noise is the dominant noise source. This noise does not include any effects of the second-stage amplifier (U2).

$$i_{BB} := 0.5 \frac{pA}{\sqrt{Hz}} \qquad \text{From data sheet}$$

$$I_{nBB} := i_{BB} \cdot \sqrt{BW_n} = 2.203 \times 10^{-12} \, A$$

$$R_1 := 2 \, k\Omega \qquad R_f := 200 \, k\Omega$$

$$R_{eq} := \frac{R_1 \cdot R_f}{R_1 + R_f} = 1.98 \times 10^3 \, \Omega$$

$$E_{n\_i1} := I_{nBB} \cdot R_{eq} = 4.363 \times 10^{-9} \, V \qquad \begin{array}{l}\text{Current noise flowing}\\\text{through feedback network}\end{array}$$

$$R_{in} := 100 \, k\Omega$$

$$E_{n\_i2} := I_{nBB} \cdot R_{in} = 220.323 \times 10^{-9} \, V \qquad \begin{array}{l}\text{Current noise flowing}\\\text{through input resistance}\end{array}$$

$$E_{n\_i} := \sqrt{E_{n\_i1}^2 + E_{n\_i2}^2} = 220.366 \times 10^{-9} \, V \qquad \begin{array}{l}\text{Total noise from}\\\text{current noise}\end{array}$$

Figure 3.11: Total noise from first-stage noise current

$$k_b := 1.38 \cdot 10^{-23} \, J/K$$

$$T_a := 298 \, K$$

$$E_{n\_r1} := \sqrt{4 \cdot k_b \cdot T_a \cdot R_{eq} \cdot BW_n} = 25.1 \times 10^{-9} \, V$$

$$E_{n\_r2} := \sqrt{4 \cdot k_b \cdot T_a \cdot R_{in} \cdot BW_n} = 178.7 \times 10^{-9} \, V$$

$$E_{n\_r} := \sqrt{E_{n\_r1}^2 + E_{n\_r2}^2} = 180.5 \times 10^{-9} \, V \quad \text{Total noise from resistor noise}$$

Figure 3.12: Total noise from first-stage resistor noise

The second-stage calculations are the same as the first-stage calculations except that the feedback network and source resistance have different values. Figure 3.14 summarizes the results. Figure 3.15 shows the calculation for the total noise from both stages. Note that the second-stage noise does not have a significant effect on the total noise. This is because the input-stage noise was amplified by a factor of 101 before combining with the second-stage noise. In general, the input-stage noise will dominate. This is especially true when the first stage is in high gain. Because the input stage will dominate, engineers will often use more expensive low-noise amplifiers for the input stage and less expensive commodity amplifiers for the output stage.

$$E_{n\_i} = 220.4 \times 10^{-9} \text{ V}$$

$$E_{n\_v} = 17.5 \times 10^{-9} \text{ V}$$

$$E_{n\_r} = 180.5 \times 10^{-9} \text{ V}$$

$$E_{n\_in} := \sqrt{E_{n\_i}^2 + E_{n\_v}^2 + E_{n\_r}^2} = 285.4 \times 10^{-9} \text{ V}$$ Total noise in the input stage (U1).

Figure 3.13: Total noise in the first stage

$$E_{n\_i} = 110.234 \times 10^{-9} \text{ V}$$

$$E_{n\_v} = 17.456 \times 10^{-9} \text{ V}$$

$$E_{n\_r} = 128.65 \times 10^{-9} \text{ V}$$

$$E_{n\_in} := \sqrt{E_{n\_i}^2 + E_{n\_v}^2 + E_{n\_r}^2} = 170.3 \times 10^{-9} \text{ V}$$ Total second stage noise from current, voltage, and resistance noise

Figure 3.14: Total noise from second-stage op-amp noise voltage

$$E_{n\_in1} := 285.4 \text{ nV}$$

$$Gain1 := 101 \quad Gain2 := 11$$

$$E_{out1} := E_{n\_in1} \cdot Gain1 \cdot Gain2 = 317.079 \times 10^{-6} \text{ V}$$ Total output from input stage noise

$$E_{n\_in2} := 170.3 \text{ nV}$$

$$E_{out2} := E_{n\_in2} \cdot Gain2 = 1.873 \times 10^{-6} \text{ V}$$ Total output from second stage noise

$$E_{n\_out\_total} := \sqrt{E_{out1}^2 + E_{out2}^2} = 317.085 \times 10^{-6} \text{ V}$$ Total output noise for example #2 including all noise sources.

Figure 3.15: Total noise from second-stage op-amp noise voltage

Chapter Summary

- The gain bandwidth of an amplifier will set the noise bandwidth if no other filters exist.
- A simple RC filter can be used to set the noise bandwidth.
- In two-stage amplifiers, the input stage will dominate.
- Hand calculations give insight where a source dominates (voltage noise, current noise, or resistor noise).
- Current noise or resistor noise may dominate for amplifiers with large sources resistances.
- Voltage noise typically dominates for amplifiers with low source resistance and low resistance feedback networks.

Questions

3.1   a. What is the noise bandwidth for the circuit below?
   b. What is the total output noise from op-amp voltage noise?
   c. What is the total output noise from op-amp current noise?
   d. What is the total output noise from op-amp resistor noise?
   e. What is the total combined output noise from all noise sources?

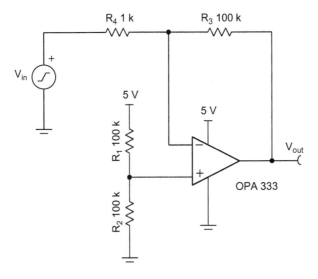

3.2   a. What is the noise bandwidth for the circuit below?
   b. What is the total output noise from op-amp voltage noise?
   c. What is the total output noise from op-amp current noise?
   d. What is the total output noise from op-amp resistor noise?
   e. What is the total combined output noise from all noise sources?

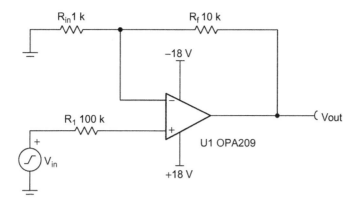

Further Reading

Hogg R.V., Tanis E.A., 1988, Probability and Statistical Inference, third ed. Macmillan, London.
Motchenbacher C.D., Connelly J.A., 1993, Low-Noise Electronic System Design, Wiley-Interscience, New York.

CHAPTER 4

Introduction to Spice Noise Analysis

In Chapter 3, we had used a hand analysis on a simple op-amp circuit. In this chapter, we use a circuit simulation package "TINA Spice" to analyze op-amp circuits. A free version of TINA Spice (TINA-TI) can be downloaded from the Texas Instruments website (http://www.ti.com, type in TINA under search). TINA Spice can do the traditional types of simulations associated with Spice packages (e.g., DC, transient, frequency domain analysis, noise analysis, and more). Furthermore, TINA-TI comes loaded with many TI analog macromodels.

In this chapter, we introduce TINA noise analysis and show how to prove that the op-amp macromodel accurately models noise. It is important to understand that some models may not properly model noise and so a simple test procedure is used to check this. This problem is solved by developing our own models using discrete noise sources and a generic op-amp.

4.1 Running a Noise Analysis in TINA Spice

In this section, we will see how to do basic noise analysis using TINA Spice. Other circuit analysis programs can do similar analysis. The circuit shown in Figure 4.1 was drawn using TINA Spice. Before running a noise analysis, it is always advisable to run a DC nodal analysis to ensure that the circuit is connected properly. If the amplifier does not have a good DC operating point, the noise analysis will not work. In the example shown, the DC output is approximately correct ($V_{out} = V_{in} \times \text{Gain} = (10\,\text{mV}) \times (-100) = -1\,\text{V}$). If the device does not operate as expected, check the power supply and ground connections.

Figure 4.2 shows how to start a noise analysis in TINA Spice. It is important to choose a frequency range that is appropriate for your application. The "diagrams" section selects the type of analysis that is done. Selecting the "Output Noise" diagram gives an output noise spectral density curve at each meter or probe. The "Total Noise" diagram gives the "integrated" total RMS noise vs. frequency.

Figure 4.3 shows the "Output Noise" and "Total Noise" plots generated using TINA noise analysis for the circuit in Figure 4.1. Note that the total noise converges to a final value of 343-μV RMS. The total noise integration will converge when a bandwidth limitation is placed on the spectral density curve. In this case, the spectral density curve bandwidth is limited by the op-amp gain bandwidth limitations. Performing the hand calculations described in Chapter 3 on this circuit will yield approximately 343-μV RMS.

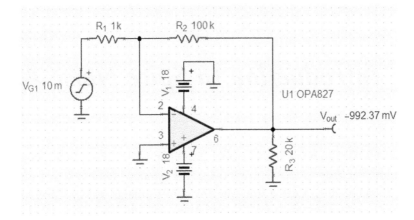

Figure 4.1: Example circuit for TINA Spice analysis

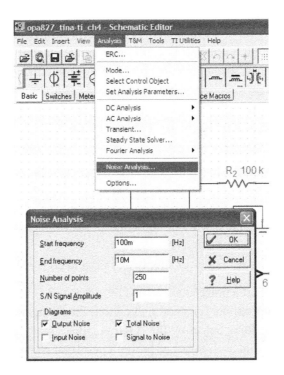

Figure 4.2: Starting TINA Spice noise analysis

4.2 Test the Op-Amp Model Noise Accuracy

The previous example assumes that the op-amp model accurately models noise. Most modern models accurately account for noise; however, some older models do not properly model noise. Figure 4.4 illustrates the test circuit used to verify the accuracy of the noise model of an op-amp.

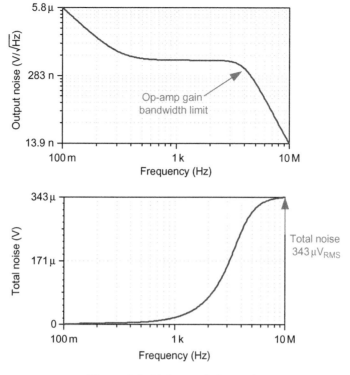

Figure 4.3: Noise analysis results

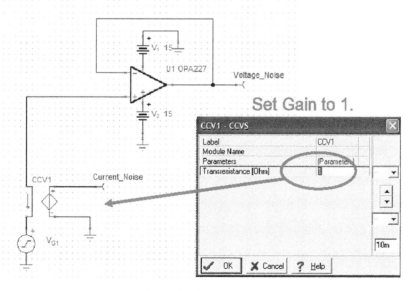

Figure 4.4: Configuration of noise test circuit (set up CCV1 gain)

CCV1 is a current-controlled voltage source that is used to convert the noise current to the noise voltage. The reason for this conversion is that the TINA "Output Noise Analysis" strictly looks at noise voltage. The gain of the CCV1 must be set to 1 as shown so that current is directly translated into voltage. The op-amp is put into a voltage follower configuration so that the input noise is reflected to the output. The two output measurement nodes "Voltage_ Noise" and "Current_Noise" are identified by TINA as nodes used to generate noise plots. The source V_{G1} is added because TINA requires an input source for noise analysis. I configure this source as sinusoidal, but this is not critical for noise analysis (see Figure 4.5).

Next a noise analysis must be done. Select "Analysis\Noise Analysis" from the pull-down menu as shown in Figure 4.6. This will bring up the Noise Analysis form. Enter the start and end frequency of interest. This frequency range is determined by the specifications of the op-amp under test. For this example, examining the specifications for the OPA227 shows that a frequency range of 0.1 Hz to 10 kHz is appropriate because this is the range that noise is specified over. Select the "Output Noise" option under "Diagrams." The "Output Noise" option will generate a spectral density curve for each measurement node (meter) in the circuit. So, when this analysis is run, we get two spectral density plots, one for the "Voltage Noise" node and one for the "Current Noise" node.

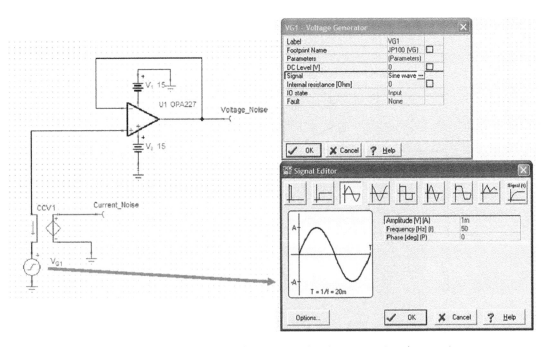

Figure 4.5: Configuration of noise test circuit (set up signal source)

The noise analysis results are shown in Figure 4.7. A few simple tricks can be used to convert these curves to a more useful format. First use "Separate Curves" under the "View" menu. Next click on the *y*-axis and select the "Logarithmic" scale. Set the lower limit and the upper limit to the appropriate range (round to powers of 10). The number of "Ticks" should be

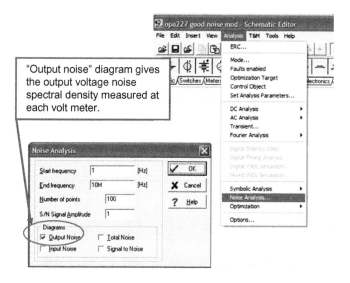

Figure 4.6: Running the "Noise Analysis" option

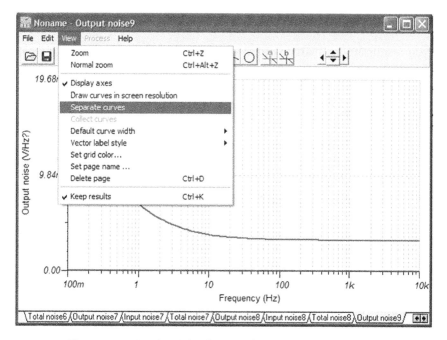

Figure 4.7: Simple tricks clean up format (separate curves)

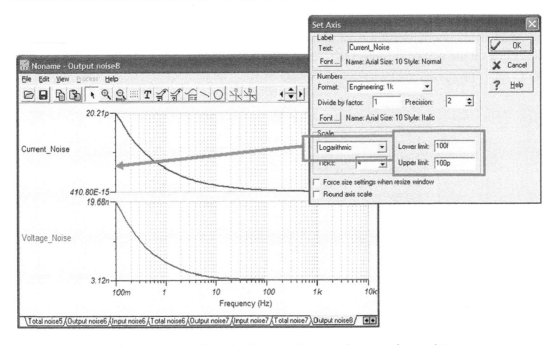

Figure 4.8: Simple tricks clean-up format (change to log scale)

adjusted to 1 + Number_of_Decades. So in this case, we have three decades (i.e., 100 f to 100 p) and so we need four ticks (see Figure 4.8).

The simulation results are compared with the OPA227 Data Sheet in Figure 4.9. Note that the two are virtually identical. Thus, the TINA-TI model for the OPA227 accurately models noise. The same procedure has been followed for the OPA627 model. Figure 4.10 shows the results for this test. The OPA627 model does not pass the test. The current noise spectral density of the model is approximately 3.5^{-21} A/rt-Hz and the specification is 2.5^{-15} A/rt-Hz. Furthermore, the voltage noise in the model shows no 1/f region. In the next section, we will build a model for this op-amp that properly models noise.

4.3 Build Your Own Noise Model

In Chapter 2, the op-amp noise model has been introduced. It consists of an op-amp, a voltage noise source, and a current noise source. We will build this noise model using discrete noise sources and a generic op-amp. These sources can be downloaded for your use from http://www.ti.com (search for TINA Noise Sources).

Figure 4.11 illustrates the circuit used to create the noise model. Note that it is in the test-mode configuration that we have used previously. We will customize this circuit to properly model the noise characteristics of the OPA627.

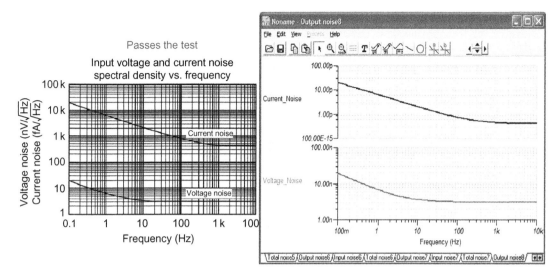

Figure 4.9: OPA227 passes the model test

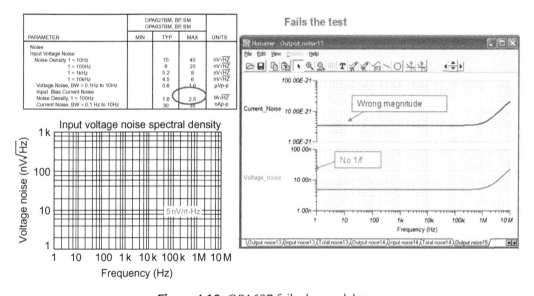

Figure 4.10: OPA627 fails the model test

First, we must configure the noise voltage source. This is done by right-clicking on the source and selecting "Enter Macro" (see Figure 4.12). Enter Macro will bring up a text editor that has the Spice macromodel listing for the source. Figure 4.13 shows the ".PARAM" information that needs to be edited to match the data sheet. Note that NLF in the Spice listing of Figure 4.13 is the noise spectral density magnitude (in nV/rt-Hz) of a point in the 1/f region. FLF in the Spice listing shown in Figure 4.13 is the frequency of the selected point.

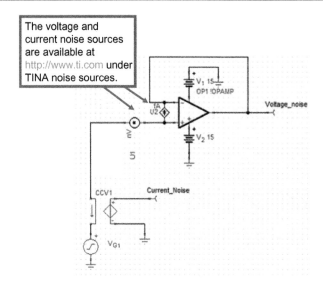

The voltage and current noise sources are available at http://www.ti.com under TINA noise sources.

Figure 4.11: Op-amp noise model using discrete noise sources

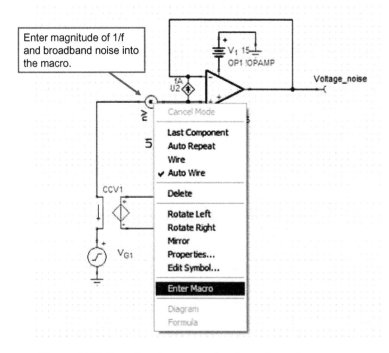

Enter magnitude of 1/f and broadband noise into the macro.

Figure 4.12: Entering the macro for the noise voltage source

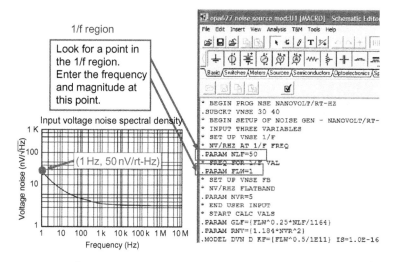

Figure 4.13: Entering the data for the 1/f region

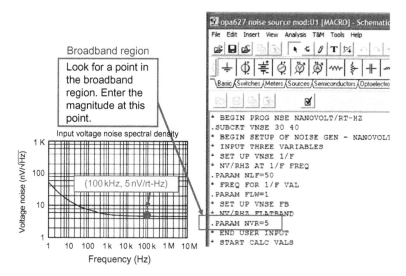

Figure 4.14: Entering the data for the broadband region

Next we need to enter the broadband noise spectral density. This is done using the NVR parameter. Note that a frequency is not required because the magnitude of the broadband noise is constant over all frequencies (see Figure 4.14). After entering the noise information, we must compile and close the Spice text editor. Press the check box and note that a "Successfully compiled" message will appear in the status bar. Under the "file" menu, select "close" to return to the schematic editor (see Figure 4.15).

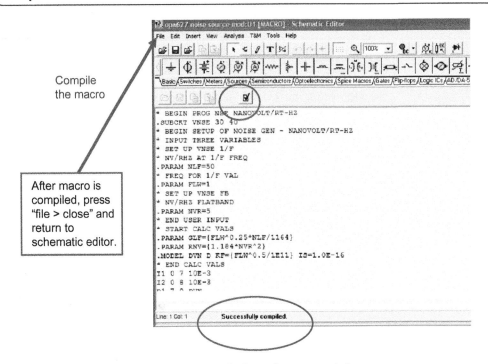

Compile the macro

After macro is compiled, press "file > close" and return to schematic editor.

Figure 4.15: Compilation of macro and close

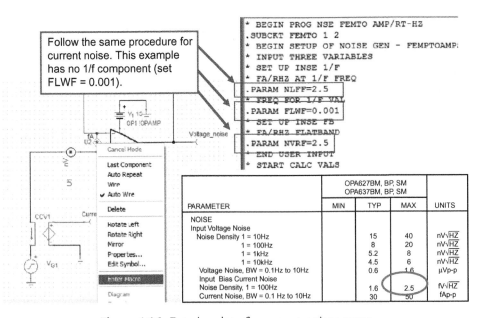

Follow the same procedure for current noise. This example has no 1/f component (set FLWF = 0.001).

Figure 4.16: Entering data for current noise source

The same procedure must be performed on the noise current source. For this example, there is no 1/f noise for the current source. In this case, the broadband and 1/f ".PARAM" are set to the same level (2.5 fA/rt-Hz). The 1/f frequency is set to some very low frequency out of the range of normal interest (i.e., 0.001 Hz) (see Figure 4.16).

Now that both noise sources have been properly configured, we must edit some AC parameters in the generic op-amp model. Specifically the open loop gain and dominant pole must be entered. The open loop gain is normally given in dB in the data sheet. Eq. (4.1) in Figure 4.17 is used to convert from dB to linear gain. Eq. (4.2) is used to calculate the dominant pole in the AOL curve. Figure 4.19 shows the calculation of the dominant pole for the OPA627. The dominant pole is shown graphically in Figure 4.18.

$$OLG = 10^{(Ndb/20)}$$ (4.1) Open loop gain in V/V

where

Ndb is the open loop gain in dB

$$Dominant\_Pole = \frac{GBW}{OLG}$$ (4.2) The low frequency pole in the open loop gain curve

where

GBW is the gain bandwidth product
OLG is the open loop gain in V/V

Figure 4.17: Calculation of the dominant pole

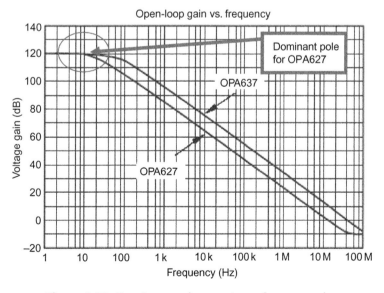

Figure 4.18: Dominant pole on gain vs. frequency plot

$$\text{OLG} = 10^{\text{Ndb}/20} = 10^{(120/20)} = 1 \cdot 10^6 \text{ V/V}$$

$$\text{Dominant\_Pole} = \frac{\text{GBW}}{\text{OLG}} = \frac{16\,\text{MHz}}{1 \cdot 10^6} = 16\,\text{Hz}$$

OPA627
data sheet

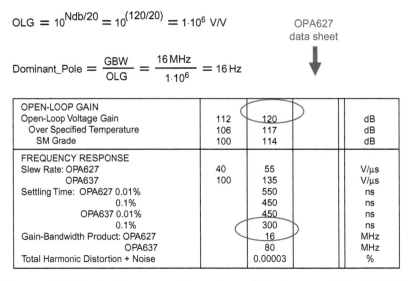

OPEN-LOOP GAIN				
Open-Loop Voltage Gain	112	120		dB
Over Specified Temperature	106	117		dB
SM Grade	100	114		dB
FREQUENCY RESPONSE				
Slew Rate: OPA627	40	55		V/µs
OPA637	100	135		V/µs
Settling Time: OPA627 0.01%		550		ns
0.1%		450		ns
OPA637 0.01%		450		ns
0.1%		300		ns
Gain-Bandwidth Product: OPA627		16		MHz
OPA637		80		MHz
Total Harmonic Distortion + Noise		0.00003		%

Figure 4.19: Finding linear open loop gain and dominant pole for OPA627

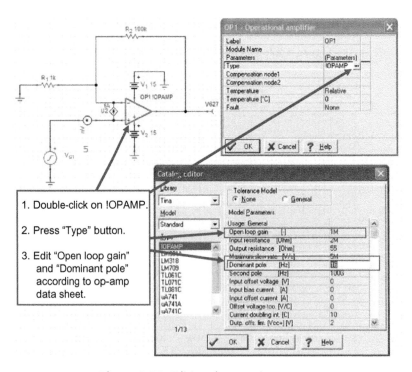

1. Double-click on !OPAMP.

2. Press "Type" button.

3. Edit "Open loop gain" and "Dominant pole" according to op-amp data sheet.

Figure 4.20: Editing the generic op-amp

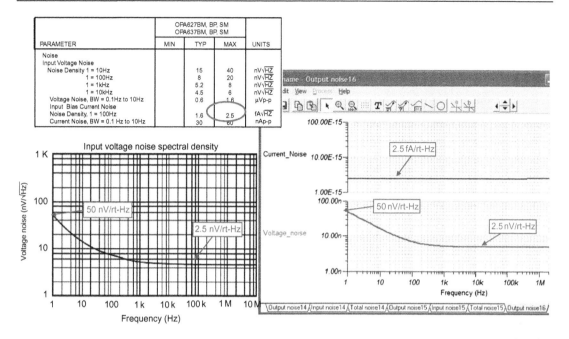

Figure 4.21: The new "hand built" model passes the model test

Next the generic op-amp model must be edited to include the open loop gain and dominant pole. This is done by double-clicking on the op-amp symbol and pressing the "Type" button. This brings up the "Catalog Editor." From the Catalog Editor, change the "Open loop gain" to match what we had calculated using the results from Figure 4.19. Figure 4.20 summarizes this procedure.

Now the op-amp noise model is complete. Figure 4.21 shows the result of running the test procedure on the model. As expected the new model matches the data sheet exactly.

4.4 Use TINA to Analyze the Circuit Given in Chapter 3

Figure 4.22 illustrates the schematic for OPA627 entered into TINA Spice. This is the circuit that has been analyzed by hand in Chapter 3.

A TINA Spice noise analysis is run by selecting "Analysis\Noise Analysis" from the pull-down menu. This will bring up the noise analysis form. Select the "Output Noise" and "Total Noise" options on the noise analysis form. The "Output Noise" option will generate a noise spectral density plot for all measurement nodes (i.e., nodes with meters). "Total Noise" will generate a plot of the integrated power spectral density curve. The total noise curve will allow us to determine the RMS output noise voltage for this circuit. Figure 4.23 shows how to run the noise analysis.

The results of the TINA noise analysis are shown in Figures 4.24 and 4.25. Figure 4.24 shows the noise spectral density at the output of the amplifier (i.e., output noise). This curve combines all the noise sources and includes the effects of noise gain and noise bandwidth. Figure 4.25 shows the total noise at the output of the amplifier for a given bandwidth. This curve was derived by integrating the power spectral density curve (i.e., the voltage spectral density squared). Note that the curve is at a constant 323-μV RMS at high frequency. This result compares very well to the RMS noise computed in Chapter 3 (i.e., the computed noise was 324 μV). Note that the noise is a constant value because of the op-amp bandwidth limitations.

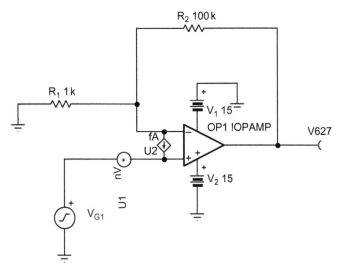

Figure 4.22: OPA627 example circuit

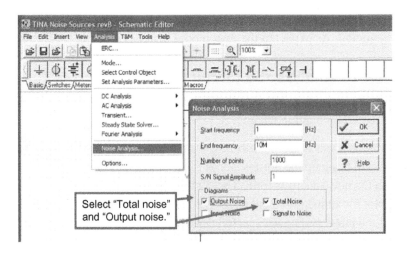

Figure 4.23: Running the noise analysis

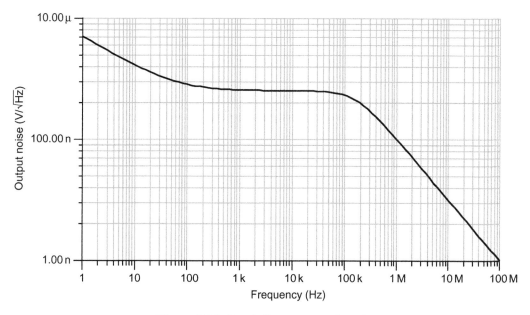

Figure 4.24: Result for output noise plot

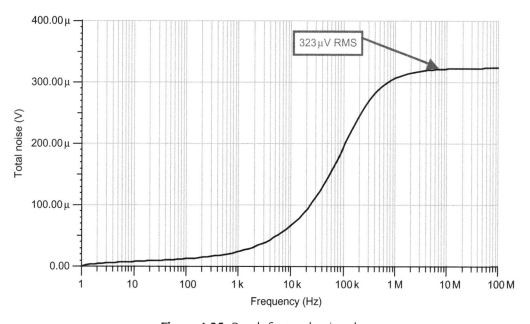

Figure 4.25: Result for total noise plot

4.5 Feedback Capacitor Simulation Example

A common op-amp circuit uses a feedback capacitor to limit the bandwidth. Limiting op-amp bandwidth will reduce noise, so a feedback capacitor is a common way to reduce noise. To understand how the feedback capacitor works, consider that a capacitor acts as a short to "high frequency" AC signals. Thus, at high frequencies the capacitor will short out the feedback resistor. When the feedback resistor is shorted, the noise gain will reduce gain to unity.

In the circuit shown in Figure 4.26, the DC noise gain is 11. The gain begins to roll off at the pole $fp = 1/(2\pi C_1 R_f)$. As the capacitive reactance decreases, the gain decreases to unity. At higher frequency the gain begins to decrease because of the op-amp gain bandwidth limitation. Figure 4.27 shows the noise gain transfer function and Figure 4.28 shows the output spectral density curve. Figure 4.29 shows the total RMS noise. Note that the region where the noise gain is unity contributes to the total noise. This region can be eliminated using an external filter (see Figure 4.30).

Figure 4.30 shows an op-amp circuit with a filter at the output. This circuit has an advantage over the feedback capacitor filter because the external filter does not have the region where the noise gain is limited to unity. Figure 4.31 compares the two filtering techniques. Note that the external filter continues to attenuate but the C_f filter attenuation stops attenuating when the gain is one (0 dB).

Figure 4.32 shows the noise spectral density of both filter types. From Figure 4.33, you can see that the external filter is more effective at reducing noise. The feedback capacitance filter

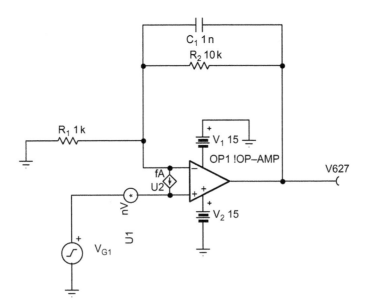

Figure 4.26: Feedback capacitor simulation example circuit

is most effective on amplifiers with high gain. Remember that the feedback capacitance will reduce the gain to unity at high frequency. The external filter is effective regardless of the gain; however, it increases the output impedance. The external filter is useful when the next stage is high impedance.

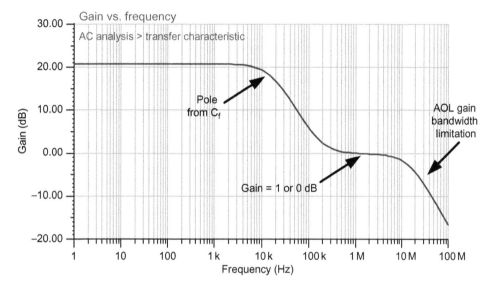

Figure 4.27: Noise gain vs. frequency for Figure 4.26

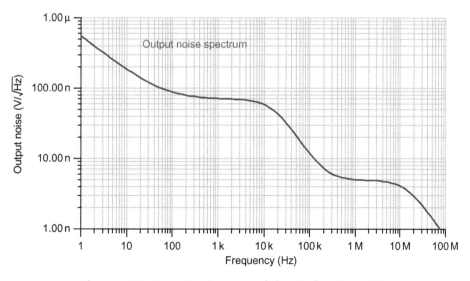

Figure 4.28: Output noise spectral density for Figure 4.26

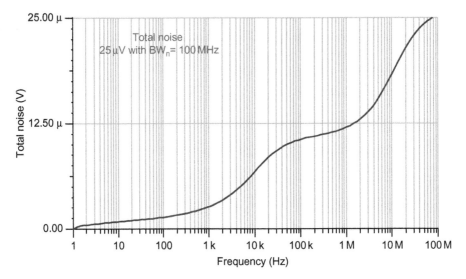

Figure 4.29: Total noise for Figure 4.26

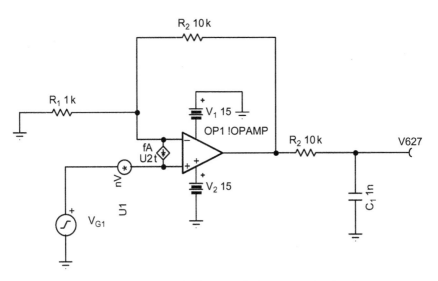

Figure 4.30: External filter simulation example circuit

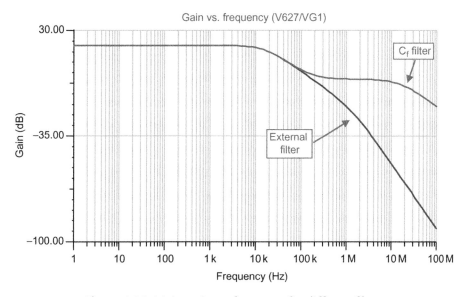

Figure 4.31: Noise gain vs. frequency for different filters

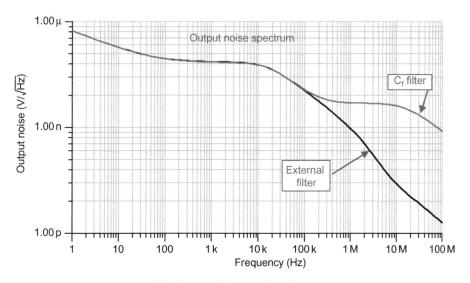

Figure 4.32: Spectral density for different filters

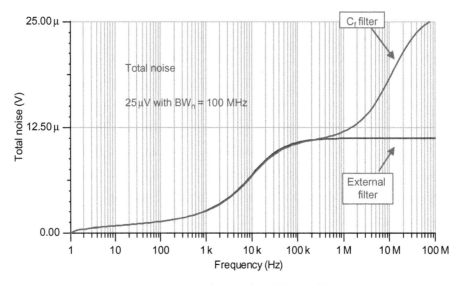

Figure 4.33: Total noise for different filters

Chapter Summary

- TINA Spice can be used to simulate the noise spectral density and total noise for op-amp circuits.
  - Verify that the model is accurate before relying on simulation results.
  - Test the DC operating point. The noise simulations will not work if the device does not have a valid DC operating point.
  - Adjust the simulation frequency range according to your application. The total noise should converge to a final value if a filter is limiting the bandwidth.
- In cases where an op-amp noise model does not exist, you can develop your own model.
  - Voltage and current noise sources generate the noise. Enter 1/f and broadband noise levels.
  - Op-amp open loop gain and dominant pole need to be entered into the generic op-amp model.

Questions

4.1   Use TINA Spice to find the spectral density and total noise for the circuit shown below.

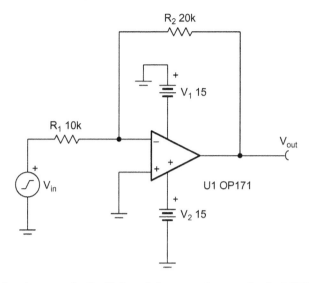

4.2   Calculate the dominant pole (in Hz) and the open loop gain (in V/V) for the OPA209.

4.3   Develop an op-amp noise model for OPA827.

Further Reading

Hogg, R.V., Tanis, E.A., 1988. Probability and Statistical Inference, third ed. Macmillan, London.

Motchenbacher, C.D., Connelly, J.A., 1993. Low-Noise Electronic System Design, Wiley-Interscience, New York.

CHAPTER 5

Introduction to Noise Measurement

In Chapter 4, we have used TINA Spice to analyze noise in operational amplifier (op-amp) circuits. The example circuit used for TINA Spice analysis is also used in the hand analysis example given in Chapter 3. The result from hand analysis and TINA Spice closely matched each other. In Chapter 5, we introduce different types of equipment used for measuring noise. Specifications and modes of operation pertinent to noise measurement are discussed. Although specific models of equipment are considered, the concepts can be applied to most equipment.

5.1 Equipment for Measuring Noise: True RMS DMM

There are three categories of test equipment used in measuring noise: true root-mean-square (RMS) meter, oscilloscope, and spectrum analyzer. The true RMS meter measures the RMS voltage for an AC signal regardless of the waveform shape. Many meters compute the RMS value by detecting the peak voltage and multiplying the peak value by 0.707. Meters using this method are not true RMS meters because they assume the wave shape is sinusoidal. A true RMS meter, however, can measure nonsinusoidal waveforms such as noise.

Many precision digital multimeters (DMMs) have true RMS capability. Typically, the meter does this by digitizing the input voltage, collecting thousands of samples, and mathematically computing the RMS value. A DMM generally has two configurations for making this measurement: "AC" and "AC + DC." In the "AC" configuration, the DMM input voltage is AC coupled to the digitizer. Thus, the DC component is stripped off. This is the preferred mode of operation for broadband noise measurements, because the result is mathematically equivalent to the standard deviation of the noise. In the "AC + DC" mode, the input signal is directly digitized and the RMS value is computed. This mode of operation should not be used for broadband noise measurements. See Figure 5.1 for a block diagram of a typical precision true RMS meter.

When using a true RMS DMM to measure noise, you must consider its specifications as well as its different modes of operation. Some DMMs have a special mode of operation optimized for making broadband noise measurements. In this mode, the DMM is a true RMS, AC coupled mode of operation that measures broadband noise from 20 Hz to 10 MHz. Also, 20-μV RMS is a typical noise floor for a precision DMM. See Table 5.1 for a summary of these specifications. Note that the noise floor can be measured by simply connecting a short across the DMM input.

Operational Amplifier Noise.

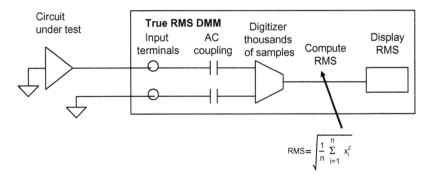

$$RMS = \sqrt{\frac{1}{n} \sum_{i=1}^{n} x_i^2}$$

Figure 5.1: Example of a typical precision true RMS DMM

Table 5.1: Summary of Typical Precision Meter Specifications

- Multiple true RMS modes read specifications to select the best mode for noise measurements

- Specified bandwidth (BW = 20 Hz to 10 MHz)

- Accuracy 0.1% for specified bandwidth

- Noise floor 20-μV RMS (on 10-mV range)

- Ranges: 10 mV, 100 mV, ..., 1000 V

5.2 Equipment for Measuring Noise: Oscilloscope

One disadvantage of using a true RMS meter for measuring noise is that the meter does not help you to know the nature of the noise. For example, the true RMS meter cannot tell the difference between noise pickup at a specific frequency and broadband noise. The oscilloscope, however, allows you to observe the time-domain noise waveform. Note that most different types of noise have a distinct waveform shape, so you can determine what type of noise dominates.

Both digital and analog oscilloscopes can be used to measure noise. Since noise is random in nature, analog oscilloscopes cannot trigger on the noise signal. Analog oscilloscopes can only trigger on repetitive waveforms. Nevertheless, analog oscilloscopes display distinctive patterns when connected to sources of noise. Figure 5.2 shows the result of a broadband measurement using an analog oscilloscope. Note that analog oscilloscopes tend to create an average or "smeared" waveform because of the phosphorescent quality of the display and the inability of the analog oscilloscope to trigger noise. One disadvantage of most standard analog oscilloscopes is that they are not able to capture low-frequency noise (1/f noise).

Digital oscilloscopes have some convenient features that help with measuring noise. Digital oscilloscopes can capture 1/f noise. Digital oscilloscopes also have the ability to mathematically compute RMS. Figure 5.3 shows the same noise source in Figure 5.2 captured using a digital oscilloscope.

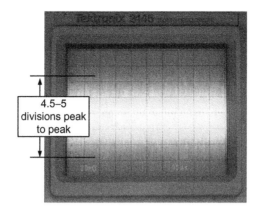

Figure 5.2: White noise on analog oscilloscope

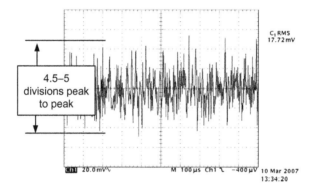

Figure 5.3: White noise on digital oscilloscope

There are some general guidelines that should be followed when using an oscilloscope for noise measurements. First of all, it is important to check the noise floor of your oscilloscope before measuring your noise signal. This can be done by connecting BNC shorting cap across the oscilloscope input or shorting the oscilloscope lead to the ground lead, if a 1 × probe is being used. This is an important consideration because the measurement range is 10 times lower when using a 1 × probe. Most good oscilloscopes have a 1 mV per division range with a 1 × oscilloscope probe or direct BNC connection, and a 10 mV per division noise floor with a 10 × probe.

Note that a direct BNC connection is preferred over a 1 × oscilloscope probe because the ground lead connection can pick up RFI/EMI interference (see Figure 5.4). One way to avoid this issue is to remove the oscilloscope probe ground lead and top cover, and use the ground on the side of the probe (see Figure 5.5). Figure 5.6 shows a BNC shorting cap.

Most oscilloscopes have a bandwidth limiting feature. To accurately measure the noise of the oscilloscopes, bandwidth must be greater than the noise bandwidth of the circuit that you are measuring. However, for best measurement results, the oscilloscope bandwidth should be limited to some value above the noise bandwidth. For example, assume that an oscilloscope has

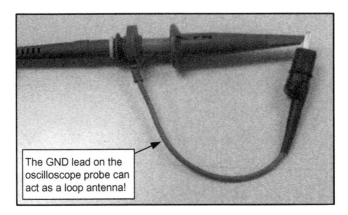

Figure 5.4: The ground lead can pick up RFI/EMI

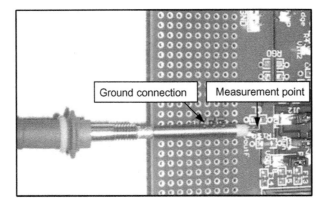

Figure 5.5: Oscilloscope probe with ground lead removed

Figure 5.6: BNC shorting cap

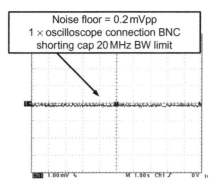

Figure 5.7: Oscilloscope noise floor with 1 × probe and bandwidth limiting

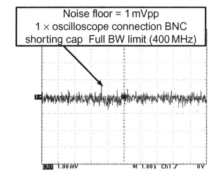

Figure 5.8: Oscilloscope noise floor with 1 × probe and without bandwidth limiting

a full bandwidth of 400 MHz, and a bandwidth of 20 MHz when the limiting feature is turned ON. If you are measuring the noise of a circuit with a noise bandwidth of 100 kHz, then it makes sense to turn on the bandwidth limiting feature. For this example, the noise floor is lower because the RFI/EMI interference outside the bandwidth of interest will be eliminated. Figures 5.7 and 5.8 show the noise floor of a typical digitizing oscilloscope with and without bandwidth limiting. Figure 5.9 shows that the noise floor is substantially higher with a 10 × probe.

The coupling mode for the oscilloscope must also be considered when making noise measurements. AC coupling should be used with broadband measurements because the noise signal generally rides upon a larger DC voltage. For example, a 1-mVpp noise signal may ride on a 2-V DC signal. Thus, in the AC coupling mode the DC signal is eliminated, allowing for the highest gain. However, note that the AC coupling mode should not be used to measure 1/f noise. This is because the bandwidth in AC coupling mode generally has a lower cutoff frequency of approximately 10 Hz. Of course, this number will vary for different models, but the point is that the lower cut frequency is too high for most 1/f noise measurements. Typically, 1/f characterization is done from 0.1 to 10 Hz. So for 1/f measurements, DC coupling with an external bandpass filter is generally used. Table 5.2 summarizes the general guidelines for noise measurements with oscilloscopes.

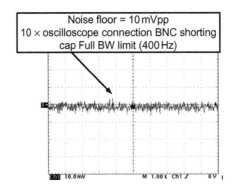

Figure 5.9: Oscilloscope noise floor with 10 × probe and without bandwidth limiting

Table 5.2: General Guidelines for Noise Measurements With Oscilloscopes

- Do not use 10 × probes for low noise measurements

- Use direct BNC connection (10 times better noise floor)

- Use BNC shorting cap to measure noise floor

- Use bandwidth limiting, if appropriate

- Use digital oscilloscope in DC coupling for 1/f noise measurements (AC coupling has a 10-Hz high–pass filter)

- Use AC coupling for broadband measurements, if necessary

5.3 Equipment for Measuring Noise: Spectrum Analyzer

The spectrum analyzer is a powerful instrument for measuring noise. Typically, the spectrum analyzer displays power (or voltage) vs. frequency similar to the noise spectral density curves. In fact, some spectrum analyzers have special modes of operation that allow the measured results to be displayed directly in spectral density units (i.e., $nV/\sqrt{Hz}$). In other cases, the results must be multiplied by a correction factor to convert the units into spectral density.

Spectrum analyzers, like oscilloscopes, may be either digital or analog. One way analog spectrum analyzers generate a spectral curve is to sweep a bandpass filter through a range of frequencies and plot the measured output of the filter. Another way is to use a superheterodyne technique where a local oscillator is swept through a range of frequencies. Digital spectral analyzers use the fast Fourier transform (FFT) (often in conjunction with a superheterodyne technique) to generate the frequency spectrum (Figure 5.10).

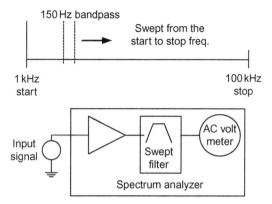

Figure 5.10: Operation of spectrum analyzer

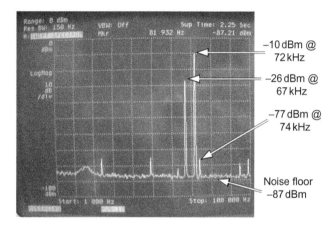

Figure 5.11: Resolution bandwidth selected for good resolution of signals

Regardless of the type of spectrum analyzer used, some key parameters need to be considered. The start and stop frequency indicates the range of frequencies that the bandpass filter is swept across. The resolution bandwidth is the width of the bandpass filter that is swept across the frequency range. Decreasing the resolution bandwidth increases the ability of the spectrum analyzer to resolve signals at discrete frequencies and causes the sweep rate to take a longer period of time. Figure 5.11 illustrates the swept filter operation. Figures 5.12 and 5.13 show two measurements with a spectrum analyzer using a different resolution bandwidth. In Figure 5.12, the resolution bandwidth is set small enough to allow proper resolving of discrete frequency components (i.e., 150 Hz). However, in Figure 5.13, the resolution bandwidth is set too wide to allow proper resolution of the frequency components (i.e., 1200 Hz).

In Figures 5.11 and 5.12, the magnitude of the spectrum is given in decibel milliwatts (dBm). This is a typical unit of measurement used by spectrum analyzers. One decibel milliwatt is

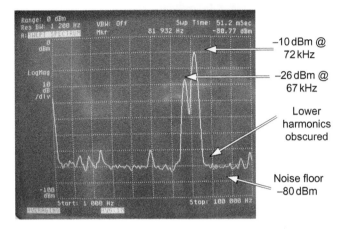

Figure 5.12: Resolution bandwidth selected for poor resolution of signals

$$N\,dBm - 10\log\frac{P}{1\,mW} \qquad (5.1)$$

where
N dBm is decibel milliwatts
P is measured power

Solve for power

$$P = \left[10^{(N\,dBm/10)}\right]\cdot(1\,mW) \qquad (5.2)$$

Power formula for resistors

$$V = \sqrt{P\cdot R} \qquad (5.3)$$

Substitute Eq. (5.2) into Eq. (5.3)

$$V = \sqrt{\left(10^{N\,dBm/10}\right)\cdot(1\,mW)\cdot R} \qquad (5.4)$$

where
R is spectrum analyzer input impedance. Some models will assume
     R = 50 Ω for both 50 Ω and 1M Ω input impedance.
N dBm is decibel milliwatts as displayed on spectrum analyzer.

Figure 5.13: Conversion of decibel milliwatts to RMS voltage

From Figure 5.13 the signal at 72 kHz has a magnitude of –10 dBm
Use Eq. (5.4) to convert –10 dBm to volts RMS.

$$V = \sqrt{\left(10^{-10\,\text{dBm}/10}\right) \cdot (1\,\text{mW}) \cdot 50\,\Omega} = 0.071\ V_{RMS}$$

Figure 5.14: Conversion of decibel milliwatts to RMS voltage

power ratio measured in decibels with reference to one milliwatt. For the spectrum analyzer in this example, the decibel milliwatt measurement also assumes a 50-Ω input impedance. For most spectrum analyzers, 50Ω is used to convert dBm to volts, even when the input impedance is selected to be 1 MΩ. Figure 5.13 gives the derivation of the formula for converting decibel milliwatts to RMS voltage. In Figure 5.14, the formula is used to compute the magnitude of the –10-dBm signal from the measurement illustrated in Figures 5.11 and 5.12.

In Figures 5.11 and 5.12, the noise floor is increased from -87 dBm to -80 dBm when the resolution bandwidth is decreased. The magnitude of the signals at 67 kHz and 72 kHz, however, do not change when the resolution bandwidth is changed. The noise floor is affected by the resolution bandwidth because it is thermal noise, and thus a wider bandwidth increases the total thermal noise. The magnitude of the signals at 67 kHz and 72 kHz is not affected by the resolution bandwidth because the signals are sinusoidal and will have a constant amplitude inside of a bandpass filter, regardless of the bandwidth. This distinction is important in noise analysis because it must be understood that discrete signals should not be included in spectral density calculations. For example, when measuring the noise spectral density of an op-amp, you may see a discrete signal at 60 Hz (power-line pickup). The 60-Hz signal should not be included in the power noise spectral density curve because it is not a spectral density signal, but rather a discrete signal.

Some spectrum analyzers display the spectral magnitude as a noise spectral density in nV/$\sqrt{\text{Hz}}$. If this feature is not available, however, the spectral magnitude can be divided by the square root of the resolution noise bandwidth to convert to spectral density. Note that a conversion factor is needed to convert the resolution bandwidth to resolution noise bandwidth. Figure 5.15 gives the equations for converting a decibel milliwatt spectrum to a spectral density. Table 5.3 gives a table of conversion factors needed to convert resolution bandwidth to noise bandwidth. Figure 5.16 shows an example where the spectrum from the example spectrum analyzer is converted to a spectral density.

Most spectrum analyzers also have an averaging feature. The averaging feature averages out the measurement variability so that the results are more repeatable. The number of averages is entered via the spectrum analyzer's front panel (typically 1–100). Figures 5.17–5.19 show the same signal measured with different levels of averaging.

$$V_{spect\_anal} = \sqrt{\left(10^{N\,dBm/10}\right) \cdot (1\,mW) \cdot R} \qquad (5.4)$$

$$V_{spect\_den} = \frac{V_{spect\_anal}}{\sqrt{K_n \cdot RBW}} \qquad (5.5)$$

where
N dBm is the noise magnitude in dBm from the spectrum analyzer.
R is the reference impedance used for the dBm calculation.
V_{spect_anal} is noise voltage measured by spectrum analyzer per resolution bandwidth.
RBW is resolution bandwidth setting on spectrum analyzer.
V_{spect_den} is spectral density in $(nV/\sqrt{Hz})$.
K_n is conversion factor that changes the resolution bandwidth to a noise bandwidth.

Figure 5.15: Formulas for conversion from dBm to spectral density

Table 5.3: Factors for Converting Resolution Bandwidth to Noise Bandwidth for Typical Spectrum Analyzers

Filter Type	Application	K_n
Four-pole sync	Most spectrum analyzers analog	1.128
Five-pole sync	Some spectrum analyzers analog	1.111
Typical FFT	FFT-based spectrum analyzers	1.056

Table from Agilent, Spectrum Analyzer Measurements and Noise Application Note 1303 (2003).

Noise floor and bandwidth are key specifications that need to be considered when using (or selecting) a spectrum analyzer. Table 5.4 lists some of the specifications for two different spectrum analyzers.

5.4 Shielding

When measuring intrinsic noise, it is important to eliminate sources of extrinsic noise. Common sources of extrinsic noise are power-line pickup (i.e., 60 Hz), monitoring noise, switching power supply noise, and wireless communication noise. Normally this is done by keeping the circuit under test inside of a shielded enclosure. The shielded enclosure is usually made of copper, iron, or aluminum. It is important that the shield be connected to system ground.

Generally, power and signal cables are connected to circuits inside the shielded enclosure via small holes in the enclosure. It is very important that the size and number of these holes be minimized. In fact, the shielding effectiveness is less of a concern than leakage through seams, joints, and holes.

Figure 5.20 shows an example of an easy-to-build and effective shielded enclosure. It is built using a steel paint can (these can be obtained at most hardware stores at nominal cost).

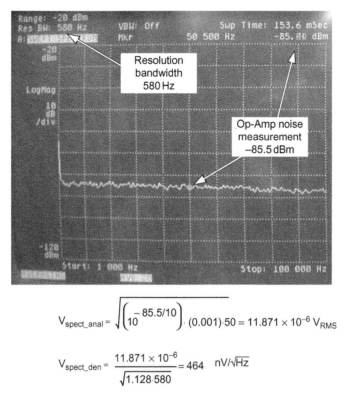

$$V_{spect\_anal} = \sqrt{\left(10^{-85.5/10}\right)} \cdot (0.001) \cdot 50 = 11.871 \times 10^{-6} \ V_{RMS}$$

$$V_{spect\_den} = \frac{11.871 \times 10^{-6}}{\sqrt{1.128 \cdot 580}} = 464 \quad nV/\sqrt{Hz}$$

Figure 5.16: Example for conversion of spectrum analyzer measurements to spectral density

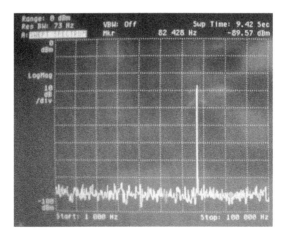

Figure 5.17: Spectrum analyzer with averaging feature turned off

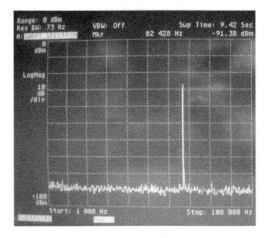

Figure 5.18: Spectrum analyzer with averaging = 2

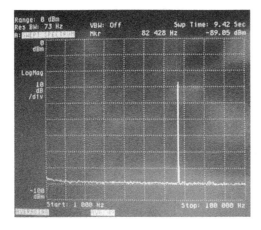

Figure 5.19: Spectrum analyzer with averaging = 49

Table 5.4: Comparing Specifications for Two Different Spectrum Analyzers

	Typical Digital Spectrum Analyzer	**Typical Analog Spectrum Analyzer**
Noise floor	20 nV/√Hz	50 nV/√Hz
Bandwidth	0.016 Hz to 120 kHz	10 Hz to 150 MHz
General comment	This is a modern digital spectrum analyzer that uses an FFT to generate the spectrum. It has very low frequency measurement capability and so it is appropriate for 1/f measurements.	This is an older generation analog spectrum analyzer that uses the superheterodyne technique to generate the spectrum. The lower cut frequency is 10 Hz and so it is not appropriate for typical op-amp 1/f measurements.

Figure 5.20: Using a steel paint can

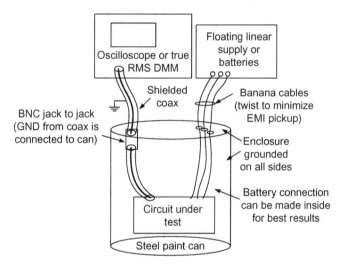

Figure 5.21: Schematic of paint can test setup

The paint can has tight seams and the lid allows easy access to the circuit under test. Note that the I/O signals are connected using shielded coaxial cables. The coaxial cable is connected to the circuit under test using a BNC jack-to-jack connector. Note that the BNC jack-to-jack connector shield is electrically connected to the paint can. The only leakage path in the enclosure is the three banana connectors used to connect power to the circuit under test. For best shielding, make sure that the paint can lid is tightly sealed. Figure 5.21 shows a schematic of the paint can test fixture.

5.5 Verify the Noise Floor

A common noise measurement goal is to measure the output noise of a low noise system or component. It is often the case that circuit output noise is too small for most standard test equipment to measure. Typically, a low-noise boost amplifier is connected between the circuit under test and the test equipment (Figure 5.22). The key to using this configuration is that the boost amplifier noise floor is lower than the circuit output noise under test, so that the

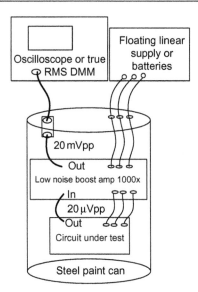

Figure 5.22: Common measurement technique

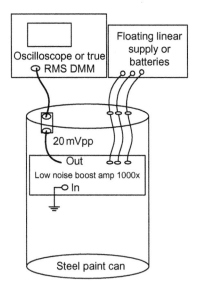

Figure 5.23: Measurement of noise floor

noise from the circuit under test dominates. As a general rule of thumb, the boost amplifier noise floor should be three times lower than the noise at the output of the circuit under test. An explanation of the theory behind this rule will be given later. *Checking the noise floor is an extremely important step that must be performed when making noise measurements.* Typically, the noise floor is measured by shorting the input of the gain block or measurement instrument. Failing to check the noise floor often leads to erroneous results (Figure 5.23).

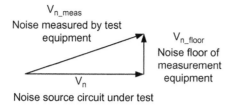

Figure 5.24: Noise adds as vectors

$$V_{n\_meas} = \sqrt{V_n^2 + V_{n\_floor}^2} \qquad (5.6)$$

The measured voltage is the vector sum of the noise from the circuit under test and the noise floor.

$$V_n = \sqrt{V_{n\_meas}^2 + V_{n\_floor}^2} \qquad (5.7)$$

Rearrange the equation to solve for the noise from the circuit under test.

Assuming the noise floor is 3 times smaller than the measured noise What measurement error is introduced by the noise floor?

$V_{n\_meas} = 1$ and $V_{n\_floor} = (1/3)$

$$V_n = \sqrt{V_{n\_meas}^2 - V_{n\_floor}^2} = \sqrt{1^2 - \left(\frac{1}{3}\right)^2} = 0.943$$

$$\text{Noise\_floor\_error} = \frac{V_{n\_meas} - V_n}{V_n} \cdot 100 = \frac{1 - 0.943}{0.943} \cdot 100 = 6\%$$

Figure 5.25: Noise floor error in percent

5.6 Account for the Noise Floor

To achieve accurate measurement results, the measurement system noise floor should be negligible in comparison to the noise level being measured. One common rule of thumb is to make sure that the noise floor is at least three times smaller than the noise signal being measured. Figure 5.24 shows how the noise output of a test circuit and the noise floor add as vectors. Figure 5.25 shows an error analysis that assumes the measured noise is three times greater than the noise floor. Using this rule of thumb, the maximum error is of the order of 6%. If you do the same calculation for a noise floor that is $10 \times$ below the measured noise, you get 0.5% error.

5.7 Measure Example Circuit #1 Using a True RMS Meter

Recall from Chapters 3 and 4 that we have analyzed a simple noninverting, op-amp circuit using the OPA627. Now we will see how this noise can be measured using a true RMS meter.

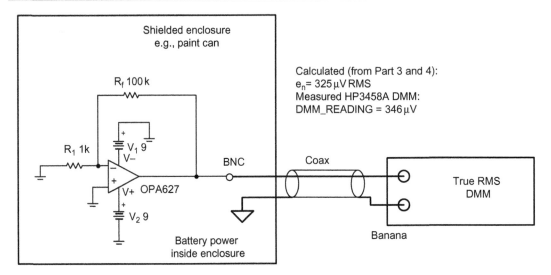

Figure 5.26: Measuring the OPA627 circuit noise with a true RMS meter

Figure 5.26 illustrates the test configuration for the OPA627. Note that the measured result for this test configuration compares well to the calculated and simulated values from Chapters 3 and 4 (i.e., calculated $325\,\mu V$ and measured $346\,\mu V$). Table 5.5 gives a step-by-step procedure for measuring noise.

5.8 Measure Example Circuit #1 Using an Oscilloscope

Figure 5.27 shows how the circuit from Chapters 3 and 4 can be measured using an oscilloscope. In the case of the oscilloscope, look at the noise waveform and estimate the peak-to-peak value. Assuming the noise has a Gaussian distribution, you can divide by 6 to get an approximation of the RMS noise (see Chapter 1 for details). The measured oscilloscope output is approximately 2.4 mVpp, so the RMS noise is 2.4 mVpp/6 = 400-μV RMS. This compares well to the calculated and simulated values from Chapters 3 and 4 (i.e., calculated $325\,\mu V$, and measured $400\,\mu V$) (Figure 5.28).

5.9 Measure Example Circuit #1 Using a Spectrum Analyzer

As we have seen in this book, the spectral density specification is an extremely important tool in noise analysis. Although most data sheets provide this information, engineers sometimes measure their own curve to verify the published data. The circuit shown in Figure 5.29 shows a simple test setup that will allow voltage noise spectral density measurements.

Note that the spectrum analyzer used for this measurement has a bandwidth of 0.064 Hz to 100 kHz. This bandwidth range allows characterization of the 1/f region and broadband region of many different amplifiers. Moreover, note that the spectrum analyzer is internally

Table 5.5: Procedure for Measuring Noise

1. Verify the noise floor of the measurement equipment (e.g., true RMS DVM). This is typically done by shorting the input to the equipment.

2. Check the specifications to ensure that the measurement equipment has appropriate bandwidth and accuracy for the proposed reading. Check the equipment specifications to see if the instrument has special modes of operation that would optimize the reading.

3. Place the circuit under test inside of a shielded enclosure. The enclosure should be connected to signal ground. Take care to minimize the size of any holes cut in the enclosure.

4. Use battery power to minimize noise, if possible. Linear power supplies are also low noise power sources. Switching power supplies are typically very noisy and will probably be the dominant source of noise if used.

5. Use shielded cables to connect the circuit under test to the measurement equipment.

6. Ensure that the circuit is functioning. In our example, the OPA627 has a typical offset voltage specified at 40 μV and the circuit gain 100, so you would expect to see an output voltage of 4 mV DC. Of course, this number will vary from device to device, but you would not expect to see an output of several volts.

7. Measure the noise using different instruments and compare the results. Using an oscilloscope and a true RMS DMM is a good approach because you can see the wave shape on the oscilloscope. The oscilloscope wave shape tells you if you have white noise, 1/f noise, 60-Hz noise pickup, or an oscillation. The oscilloscope will also give you a rough idea of the peak-to-peak noise level. The true RMS DMM, however, does not give information about the type of noise, but it does give an accurate RMS noise result. A spectrum analyzer is also a great tool in noise analysis because it highlights any problems at discrete frequencies (i.e., noise pickup or noise peaking).

8. Compare the measured result to a calculated and simulated result, if possible. In general, it is possible to get good agreement between calculated and measured results.

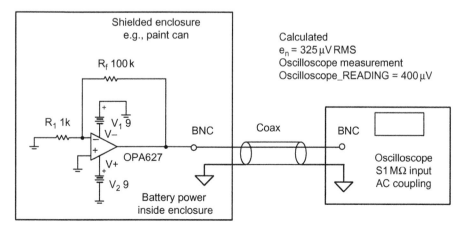

Figure 5.27: Measuring the OPA627 circuit noise with an oscilloscope

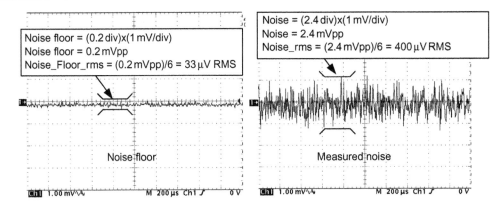

Figure 5.28: Results on oscilloscope

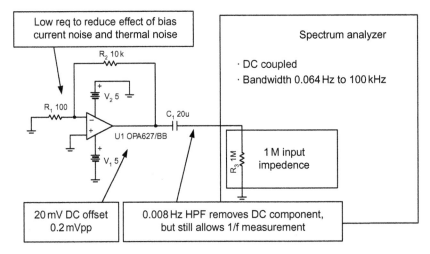

Figure 5.29: Circuit for measuring the noise spectral density of an op-amp

configured in DC-coupling mode. The instrument is not configured in AC coupling mode because its lower cut frequency is 1 Hz, and this would not allow for good 1/f readings. Nevertheless, it is desirable to AC couple the op-amp circuit to the spectrum analyzer because the DC offset is large in comparison to the noise. Thus, the op-amp circuit is AC coupled using the external coupling capacitor C_1 in conjunction with the input impedance of the spectrum analyzer R_3. The lower cut frequency of this circuit is 0.008 Hz (this will not interfere with our 1/f measurements because the spectrum analyzer's minimum frequency is 0.064 Hz). Note that C_1 is actually multiple ceramic capacitors in parallel (electrolytic and tantalum capacitors are not recommended for this application).

Another consideration with the amplifier configuration in Figure 5.29 is the value of feedback network. Remember from Chapter 2 that the parallel combination of R_1 and R_2 ($R_{eq} = R_1 \| R_2$)

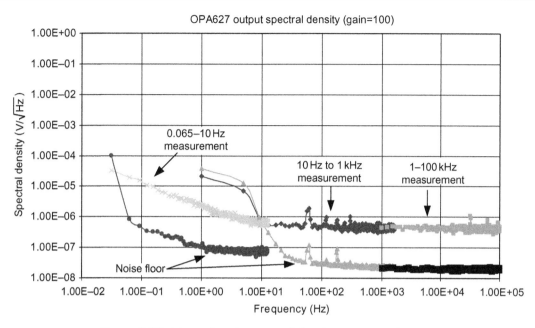

Figure 5.30: Measuring the spectral density over several ranges

is used to compute the thermal noise and the bias current noise. The value of this resistance should be minimized so that the measured noise is only from the op-amp voltage noise (i.e., the effect of bias current noise and resistor thermal noise is negligible).

As with all noise measurements, verify that the noise floor of the spectrum analyzer is lower than the op-amp circuit. For the example shown in Figure 5.29, the amplifier is configured for a gain of 100 to increase the output noise above the noise floor of the spectrum analyzer. Keep in mind that this configuration will limit the high-frequency bandwidth (i.e., Bandwidth = Gain Bandwidth Product/Gain = 16 MHz/100 = 160 kHz). Thus, the noise spectral density curve will roll off at a lower frequency. The example shown in Figure 5.29 is not affected by this issue because the high frequency roll-off occurs outside of the spectrum analyzer's bandwidth (i.e., the noise rolls off at 160 kHz and the spectrum analyzer's maximum bandwidth is 100 kHz).

Figure 5.30 shows the result of the spectrum analyzer measurement. Note that the data was collected over several different frequency ranges (i.e., 0.064–10 Hz, 10 Hz to 1 kHz, and 1–100 kHz). This is because the spectrum analyzer in this example uses a linear frequency sweep to collect data. For example, if a data point is taken every 0.1 Hz, then the resolution is poor at low frequencies and excessive at high frequencies. Also using a small resolution over a wide frequency range would require an excessive number of data points (e.g., 0.1-Hz resolution and 100-kHz bandwidth would require 1×10^6 points). However, if you change the resolution for different frequency runs, you can get good resolution in each range without

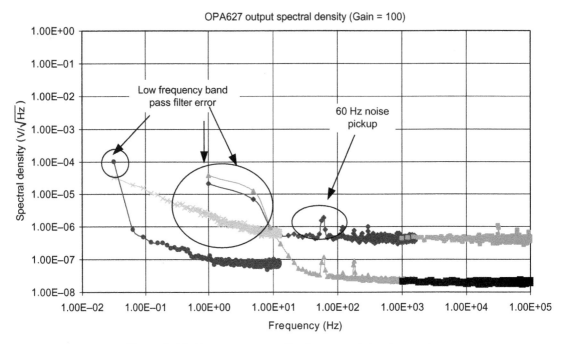

Figure 5.31: Common anomalies in spectral density results

using an excessive number of data points. For example, the resolution from 0.064 Hz to 10 Hz could be set to 0.01 Hz, while the resolution from 1 kHz to 100 kHz could be set to 100 Hz.

Figure 5.30 highlights some common anomalies in the spectrum analyzer results. The first anomaly is noise pickup from an external source. Specifically, this example shows noise pickup at 60 Hz and 120 Hz. It is also common to see pickup from the local oscillator in the spectrum analyzer. Under ideal circumstances the noise pickup would be minimized by shielding; however, in practice this pickup is often unavoidable. The key is to identify if the noise "spike" in the spectrum is the result of noise pickup, or is it the part of the intrinsic noise spectral density of the device.

Another common anomaly seen in the spectral density plot shown in Figure 5.31 is the relatively large error that occurs at the minimum frequency for a given measurement range. To better understand this error, consider that the spectrum is measured by sweeping a bandpass filter across the entire spectrum. For example, assume the range is 1 Hz to 1 kHz and the resolution bandwidth of the bandpass filter is 1 Hz. For this frequency range, the bandpass filter is relatively narrow at the high frequencies, and relatively wide at the low frequencies.

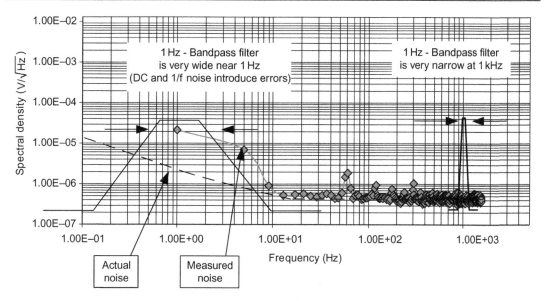

Figure 5.32: Measurements at the minimum frequency include errors

Now consider that at the low frequencies, the skirts of the bandpass filter will pick up large errors from 1/f noise. Figure 5.32 graphically illustrates this error.

Correcting the errors requires understanding of the various measurement anomalies. For example, by measuring the data over several ranges and discarding a few points on the low end of the frequency range, you can get more accurate results. In our example, the "0.0625 Hz to 10 Hz" range overlaps the "10 Hz to 1 kHz" range. The "10 Hz to 1 kHz" range contains some erroneous data below 10 Hz. This erroneous data is discarded. Noise pickup (e.g., 60 Hz pickup) can be omitted from spectral density measurements because it is not part of the intrinsic op-amp noise.

Figure 5.33 shows the noise spectral density curve from our example measurement with the anomalous readings eliminated. The data in Figure 5.33 was also divided by the test circuit gain so that the spectral density is referred to the input of the op-amp. Finally, the data was averaged.

Comparison of the spectral density measurement for the OPA627 with the data-sheet curve shows an interesting result. The measured and data-sheet specification for broadband noise match closely, but the 1/f noise measurement is significantly different from the specification. Actually, the deviation of the 1/f noise from the specification is not a very unusual result. In Chapter 6, we will discuss this topic in detail (Figure 5.34).

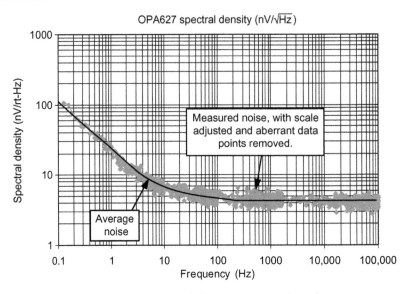

Figure 5.33: Spectral density measured results

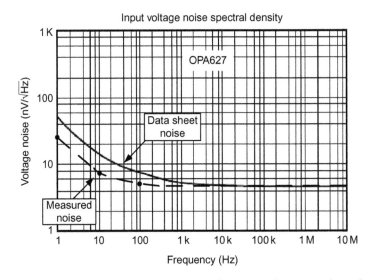

Figure 5.34: Comparison of the measured spectral density to data sheet

5.10 Measure Low Frequency Noise for the OPA227

Many data sheets have a specification for the peak-to-peak noise from 0.1 Hz to 10 Hz. This effectively gives an idea of the op-amp's low frequency (i.e., 1/f noise). In some cases, this is given as an oscilloscope waveform. In other cases, this is listed in the specification table. Figure 5.35 shows one effective way of measuring the 0.1–10 Hz noise. This circuit

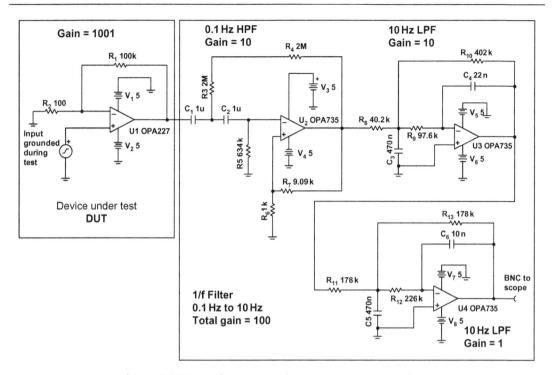

Figure 5.35: Low-frequency noise measurement test circuit

uses a second-order 0.1-Hz high-pass in series with fourth-order 10-Hz low-pass filter. This circuit also provides a gain of 100. The device under test (OPA227) is placed in a high-gain configuration (noise gain = 1001) because the expected 1/f noise is very small and must be amplified to a range where standard test equipment can be used to measure it. Note that the total gain for the circuit in Figure 6.9 is 100,100 (i.e., 100 × 1001). Thus, the output is divided by 100,100 to refer the signal to the input.

The measured output noise from the circuit shown in Figure 5.36 is shown in Figure 5.37. Figure 5.37 also shows a graph taken from the OPA227 product data sheet. The range of the measured result can be referred to the op-amp's input by dividing by the total gain (i.e., divide by 5 mV/100,100 = 50 nV). Note that the product data sheet curve compares well to the data sheet curve as expected.

5.11 Offset Temperature Drift vs. 1/f Noise in Low-Frequency Noise Measurement

One problem in measuring the 1/f noise of an amplifier is that it is often difficult to distinguish 1/f noise from offset drift with temperature. Note that ambient temperature fluctuates ±3 °C in a typical lab environment. Air turbulence around a device can cause

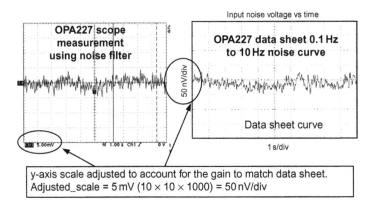

Figure 5.36: Results for low-frequency noise measurement test circuit

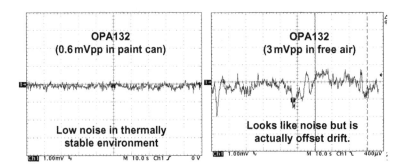

Figure 5.37: OPA132, lab environment vs. thermally stable environment

low-frequency variations that are in the offset that look similar to 1/f noise. Figure 5.37 compares the output of an OPA132 in a thermally stable environment with the same circuit in a typical lab environment. Ambient temperature drift in a lab environment can be 6C. The OPA132 specified maximum offset drift is $10\,\mu V/C$. In this example the offset drift due to ambient temperature variations is $60\,\mu V$ ($6C \times 10\,\mu V/C = 60\,\mu V$). The amplifier in Figure 6.11 is in a gain of 100, so the output drifts approximately $6\,mV$ (i.e., $(60\,\mu V)(100) = 6\,mV$).

One way to separate the effects of offset drift from 1/f noise is to place the device under test in a thermally stable environment. This environment must keep the device at constant temperature (i.e., $\pm 0.1\,°C$) throughout the measurement. Also temperature gradients should be reduced as much as possible. A simple way to achieve this goal is to fill the paint can with an electrically inert fluid and submerge the entire assembly during test. Heat-transfer fluorinated fluids are commonly used for this type of testing because of their high electrical resistance and high thermal resistance. Also, they are biologically inert and nontoxic.

Chapter Summary

- A true RMS digital multimeter can be used to measure noise.
  - Some true RMS meters have a very good noise floor.
  - A disadvantage is that you cannot distinguish different noise sources (e.g., intrinsic vs. extrinsic).
- An oscilloscope is a common instrument used to measure noise.
  - It shows waveform shape vs. time.
  - Some oscilloscopes have math functions for RMS readings.
- For oscilloscope measurements, following some guidelines can improve accuracy.
  - Using 1 × probes.
  - Using bandwidth limiting, if possible.
- A spectrum analyzer can be used to measure noise in the frequency domain.
- Increasing the spectrum analyzer resolution bandwidth will allow you to resolve spectral density changes more accurately, but will increase measurement time.
- Increasing the spectrum analyzer averaging will improve the accuracy of your measurement, but will increase measurement time.
- It is important to shield all measurements to minimize extrinsic noise pickup.
- Always verify the noise floor of the test equipment you are using.
- Ideally the noise floor should be significantly below the noise being measured. If this is not the case, subtract the noise floor using Eq. (5.7).

Questions

5.1 A spectrum analyzer reads −90 dBm. What is the spectral density? Assume 50-Ω input impedance, 1 kHz resolution bandwidth, and a four-pole sync.

5.2 The output noise of an op-amp is measured with an oscilloscope. The oscilloscope reads 1.2-mV RMS. The oscilloscope noise floor is measured to be 0.2-mV RMS. What is the actual op-amp output noise?

Further Reading

Agilent. Spectrum Analyzer Measurements and Noise Application Note 1303 2003. (http://www.agilent.com)

Hogg R.V., Tanis E.A., 1988, Probability and Statistical Inference, third ed. Macmillan, New York

Motchenbacher C.D., Connelly J.A., 1993, Low-Noise Electronic System Design, Wiley-Interscience Publication, New York

Ott H.W., 1988, Noise Reduction Techniques in Electronic Systems, second ed. John Wiley & Sons, New York
http://www.solvaysolexis.com/

CHAPTER 6

Noise Inside the Amplifier

This chapter discusses the fundamental physical relationships that determine the intrinsic noise of an operational amplifier (op-amp). Board- and system-level designers will gain insight into performance tradeoffs made by integrated circuit designers between noise and other op-amp parameters. Also, engineers will learn how to estimate worst-case noise based on typical data sheet specifications at room temperature and overtemperature. It should be emphasized that these rules of thumb are intended to cover the majority of applications; however, there will be some special cases that are exceptions to the rules.

6.1 Five Rules of Thumb for Worst-Case Noise Analysis and Design

Most op-amp data sheets list only a typical value for op-amp noise, with no information regarding temperature drift of noise. Board- and system-level designers would like to have a method for estimating maximum noise based on the typical value. Furthermore, it would be useful to estimate noise drift with temperature. There are some fundamental noise relationships for transistors that can help to make these estimates. However, to make use of these relationships precisely, some knowledge of the internal topology is required (e.g., biasing configuration and transistor type). Nevertheless, it is possible, if we consider the worst-case configuration, to make some general statements that cover the majority of configurations. This section of the chapter summarizes five basic rules of thumb for worst-case noise analysis and design. Section 6.2 gives detailed mathematics behind these rules of thumb.

Rule of thumb #1: Broadband voltage noise is very insensitive to semiconductor process changes. This is because op-amp noise is generally a function of op-amp power supply quiescent current (I_q). Typically, bias current is relatively constant from device to device. Alternatively, for some designs, noise could be dominated by the thermal noise of the input electrostatic discharge (ESD) protection resistors. With this in mind, it is unlikely that the broadband noise will change more than 10% from the typical value. In fact, this variation typically is less than 10% for most low-noise devices. See example in Figure 6.1.

Broadband current noise is more sensitive than voltage noise (for bipolar processes). This is because current noise is related to base current, which is set by the transistor current gain (β). Variation of base current can be as large as four times the typical value (i.e., a 4:1 ratio).

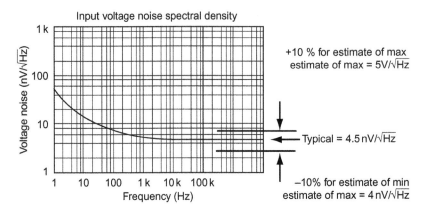

Figure 6.1: Maximum room temperature broadband noise

Broadband current noise spectral density is proportional to the square root of base current variation, so the current noise variation should be less than ± 2 times the typical value.

Rule of thumb #2: Op-amp noise increases with temperature. For many biasing schemes (e.g., proportional to absolute temperature, PTAT), the noise will increase proportionately to the square root of absolute temperature, and so the change in noise over the extended industrial temperature range is relatively small (i.e., 15% for 25–125 °C). It is possible, however, for some biasing schemes (i.e., zero-TC) to generate noise that is proportional to absolute temperature. For this worst-case scenario, noise changes 33% over the same temperature range. Figure 6.2 illustrates this graphically.

Rule of thumb #3: 1/f noise (i.e., flicker noise) is highly process dependent. This is because 1/f noise is related to defects in the crystalline structure that are created during the fabrication process. Thus, as long as the semiconductor process is well controlled, the level of 1/f noise should not shift substantially. A fabrication issue or a process change can substantially alter the 1/f noise. In cases where the device data sheet gives maximum value for 1/f noise, either the process is monitored or the device is measured at final test. If a data sheet maximum value is not given for 1/f noise, the factor of three variations is an estimate for worst case— assuming that the process control is not optimized for 1/f noise reduction. See Figure 6.3.

Rule of thumb #4: Board- and system-level designers need to understand that op-amp power supply quiescent current (I_q) and broadband noise are inversely related. Strictly speaking, noise is related to the biasing of the op-amp's input differential stage. However, since this information is not normally published, we can assume that Iq is proportional to the differential stage bias. This assumption is most accurate with low-noise amplifiers.

Typically, broadband noise is inversely proportional to the square root of Iq. However, this relationship can vary for different biasing schemes. This rule of thumb should help

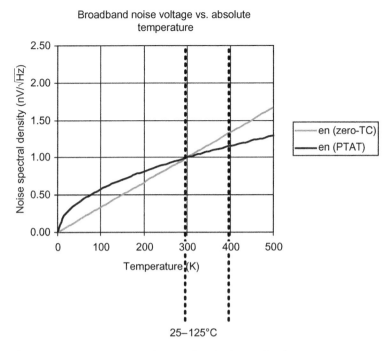

Figure 6.2: Worst-case and typical variations of noise vs. temperature

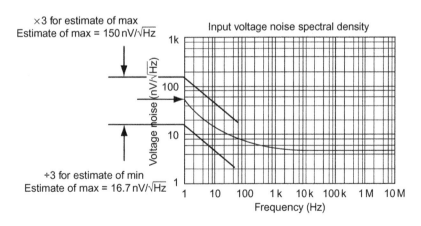

Figure 6.3: Worst-case estimate for 1/f noise

board- and system-level designers better understand the tradeoff between Iq and noise. For example, a designer should not expect amplifiers with extremely low I_q to also have low noise. Figure 6.4 illustrates this relationship graphically.

Rule of thumb #5: Field-effect transistor (FET) op-amps have inherently low current noise. This deals with the difference between bipolar and FET transistors and noise. This is because the input

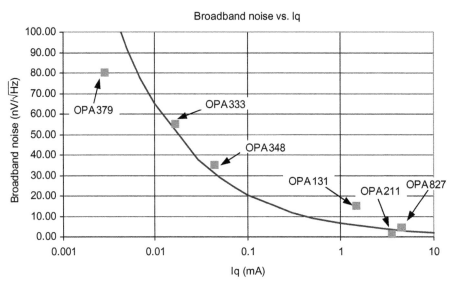

Figure 6.4: Iq vs. broadband noise

Table 6.1: Comparing CMOS With Bipolar for Voltage and Current Noise

Op-amp	Type	Iq (mA)	i_n (fA/$\sqrt{Hz}$)	e_n (nV/$\sqrt{Hz}$)
OPA277	Bipolar	0.79	200	8
OPA211	Bipolar	3.6	1500	1.1
OPA227	Bipolar	3.7	400	3
OPA348	CMOS	0.045	4	35
OPA364	CMOS	1.1	0.6	17
OPA338	CMOS	0.53	0.6	26

gate current of a FET is substantially smaller than the input base current of a bipolar amplifier. Conversely, bipolar amplifiers tend to have lower voltage noise for a given value of bias current (i.e., collector or drain current on the input stage). See several examples in Table 6.1.

6.2 Detailed Mathematics for Bipolar Noise

Figure 6.5 illustrates the schematic of the bipolar transistor noise model. The fundamental noise relationships for bipolar transistors are given in Figure 6.6 (Eqs. (6.1)–(6.3)). In this section, we will manipulate these equations to show the fundamental relationships that are the basis for these rules of thumb.

6.2.1 Analysis Using Eq. (6.1): Bipolar Thermal Noise

Eq. (6.1) represents the physical resistance thermal noise in the base of a bipolar transistor. In an integrated circuit op-amp, this resistor often is from an ESD protection circuit in series

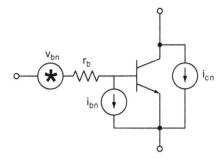

Figure 6.5: Bipolar transistor noise model

$$v_{bn}^2 = 4 \cdot kT \cdot r_b \cdot \Delta f \qquad (6.1)$$

where
v_{bn} is thermal noise at base of bipolar transistor from physical resistance
k is Boltzmann's constant (1.38×10^{-23} J/K)
T is temperature in Kelvin
r_b is physical resistance in base of transistor
Δf is noise bandwidth

$$i_{cn}^2 = 2 \cdot q \cdot I_C \cdot \Delta f \qquad (6.2)$$

where
i_{cn} is collector shot noise current
q is electron charge ($1.6 \cdot 10^{-19}$ coulomb)
I_C is DC collector current

$$i_{bn}^2 = 2 \cdot q \cdot I_B \cdot \Delta f + K_1 \cdot \frac{I_B^a}{f^b} \cdot \Delta f + K_2 \cdot \frac{I_B^c}{1 + \left(\dfrac{f}{f_c}\right)} \cdot \Delta f \qquad (6.3)$$

Shot noise   Flicker noise   Burst noise

where
i_{bn} is base current noise
I_B is DC base current
K_1 is semiconductor process dependent constant for flicker noise
a is a constant between 0.5 and 2.0
b is a constant about unity

Figure 6.6: Fundamental bipolar noise relationships

with the base of the differential input stage. See r_b from Figure 6.5. In some cases this noise dominates. For most integrated circuit processes, it is reasonable to assume ±20% tolerance for this resistance. Figure 6.7 shows that a 20% variation with input resistance corresponds to 10% variation in noise.

$$v_{bn}^2 = 4 \cdot kT \cdot r_b \cdot \Delta f$$

For a 20% variation in r_b v_{bn} will vary

$$\%\_variation\_v_{bn} = \frac{\sqrt{4 \cdot kT \cdot 1.2 \cdot r_b \cdot \Delta f} - \sqrt{4 \cdot kT \cdot r_b \cdot \Delta f}}{\sqrt{4 \cdot kT \cdot r_b \cdot \Delta f}} \cdot 100$$

$$\%\_variation\_v_{bn} = \frac{(\sqrt{1.2} - \sqrt{1}) \cdot \sqrt{4 \cdot kT \cdot r_b \cdot \Delta f}}{\sqrt{4 \cdot kT \cdot r_b \cdot \Delta f}} \cdot 100$$

$$\%\_variation\_v_{bn} = 9.5\%$$

Figure 6.7: Thermal noise tolerance

$$i_{cn}^2 = 2 \cdot q \cdot I_c \cdot \Delta f$$

In terms of voltage referred to input

$$v_{cn\_RTI}^2 = \frac{1}{g_m^2} \cdot \left(2 \cdot q \cdot I_c \cdot \Delta f\right)$$

Substitute g_m for bipolar

$$g_m = \frac{I_c}{V_t} \qquad g_m = \frac{I_c}{\frac{k \cdot T}{q}} \qquad \frac{1}{g_m} = \frac{k \cdot T}{q \cdot I_c}$$

Collector shot noise in voltage format

$$v_{cn\_RTI}^2 = \left(\frac{k \cdot T}{q \cdot I_c}\right)^2 \cdot \left(2 \cdot q \cdot I_c \cdot \Delta f\right)$$

where V_{cn_RTI} is the collector current noise
referred to the inputs as a voltage noise.

Figure 6.8: Conversion of current noise to voltage noise

6.2.2 Analysis Using Eq. (6.2): Bipolar Collector Shot Noise

Eq. (6.2) gives the relationship for a bipolar transistor collector shot noise. To better understand this relationship, it helps to convert it to a voltage noise v_{cn_rti} (see Figure 6.8). Further simplifications can be done to the formula if the biasing scheme for the input stage is known. There are two types of biasing schemes for op-amp input stages. One scheme forces the collector current to be proportional to absolute temperature (PTAT). For a PTAT biasing scheme, the collector current can be represented as a constant multiplied by absolute

Collector shot noise in voltage format

$$v_{cn\_RTI}^2 = \left(\frac{k \cdot T}{q \cdot I_c}\right)^2 \cdot \left(2 \cdot q \cdot I_c \cdot \Delta f\right)$$

Assuming PTAT $I_c = \alpha \cdot T \cdot I_c$

$$v_{cn\_RTI}^2 = \left[\frac{k \cdot T}{q \cdot \left(\alpha \cdot T \cdot I_c\right)}\right]^2 \cdot 2 \cdot q \cdot \left(\alpha \cdot T \cdot I_c\right) \cdot \Delta f$$

Combine all constants into K_a

$$v_{cn\_RTI}^2 = K_a \left(\frac{1}{I_c}\right)^2 \cdot \left(T \cdot I_c\right) \cdot \Delta f$$

$$v_{cn\_RTI}^2 = \sqrt{\frac{K_a T \cdot \Delta f}{I_c}}$$

Thus, collector shot noise voltage refered to the input is directly proportional to $\sqrt{T}$ and inversely proportional to $\sqrt{I_c}$ for a PTAT bias.

Figure 6.9: Collector noise voltage for PTAT bias

temperature. Figure 6.9 shows a simplification of the v_{cn_rti} equation based on a PTAT biasing scheme. The key result is that the noise is directly proportional to the square root of temperature and inversely proportional to the square root of I_c. This important result illustrates why low-noise amplifiers always have high I_q. This is the basis of the fourth rule of thumb. The result also shows that the op-amp noise increases with increase in temperature. This is the basis of the second rule of thumb.

Op-amp input stages also are biased in a "zero-TC" configuration where the collector current bias does not drift with temperature. Figure 6.10 shows a simplification of the v_{cn} equation based on a zero-TC bias configuration. The key result is that the noise is directly proportional to temperature and inversely proportional to the square root of I_c. The zero-TC configuration has a disadvantage over the PTAT method because it is more sensitive to changes in temperature. Note that in the second rule of thumb, this is the worst-case graph.

The result from Figures 6.9 and 6.10 can be used to determine how much noise changes when I_c is modified. In both cases, noise is inversely proportional to the square root of I_c. In an integrated circuit op-amp design, the differential input stage typically dominates the noise. Unfortunately, the data sheet does not give information about the biasing of this amplifier. To get a rough estimate, you can assume that the change in I_c is proportional to the change in quiescent current (Iq). In general, the input stage biasing is better controlled than Iq, so this

Assuming zero-TC Ic is constant over temperature

$$V_{cn\_rti} = \left(\frac{k \cdot T}{q \cdot I_c} \right)^2 \cdot \left(2 \cdot q \cdot I_c \cdot \Delta f \right)$$

Combine all constants into K_b

$$V_{cn\_rti} = K_b \left(\frac{T}{I_c} \right)^2 \cdot \left(I_c \cdot \Delta f \right)$$

$$V_{cn\_rti} = T \sqrt{\frac{K_b \Delta f}{I_c}}$$

Thus, collector shot noise voltage is directly proportional to T and inversely proportional to $\sqrt{I_c}$ for a zero-TC bias.

Figure 6.10: Collector noise voltage for zero-TC bias

Estimate worst case noise based on I_q variation

$$v_{n\_worst\_case} = v_{n\_typical} \cdot \sqrt{\frac{I_{q\_worst\_case}}{I_{q\_typical}}}$$

$$v_{n\_worst\_case} = 8 \frac{nV}{\sqrt{Hz}} \sqrt{\frac{825\ \mu A}{790\ \mu A}} = 8.2 \cdot \frac{nV}{\sqrt{Hz}}$$

Figure 6.11: Worst-case noise based on Iq variation

is a conservative estimate. Figure 6.11 shows an estimate worst-case noise for the OPA277. Note that in this case, the variation in Iq has little effect on noise. For most practical designs, this variation will be less than 10%. Note that the first rule of thumb is based on the fact that both thermal noise variations and shot noise variations (I_c variations) will be less than 10%.

6.2.3 Analysis Using Eq. (6.3): Bipolar Base Shot and Flicker Noise

Eq. (6.3) describes bipolar transistor base shot and flicker noise. This noise source is analogous to the current noise in an op-amp. This current noise source also converts to a voltage noise as is shown in Figure 6.12. Analyzing the PTAT and zero-TC bias configuration is not as straightforward as in the collector current shot noise case. This is because the bias methods are designed to control the collector current and the relationship does not follow for the base current. For example, a device with zero-TC collector current will not have zero-TC base current because bipolar current gain changes with temperature.

$$i_{bn}^2 = K_1 \cdot \frac{I_B^a}{f} \cdot \Delta f$$

where K_1 is the flicker constant and is process dependent.
Flicker current will translate to voltage through input resistance.

$$v_{bn} = r_{b} \cdot \sqrt{K_1 \cdot \frac{I_B^a}{f} \cdot \Delta f}$$

Figure 6.12: Flicker noise voltage relationship

The shot noise component in Eq. (6.3) is responsible for broadband current noise. Also note that the noise current is proportional to the square root of Ib. This is why broadband current noise is more sensitive than broadband voltage noise. Variations in Ib are caused by the current gain (β) of the transistor.

Note that the shot noise component is in the same form as Eq. (6.2). So the analysis is the same, except that the temperature coefficient of the base current is difficult to predict. Thus, for the sake of simplicity, we will not include temperature information for the i_b shot noise.

The flicker noise component is converted into a voltage noise in Figure 6.12. Note that flicker noise increases with increase in temperature and decreases with decrease in I_c. Flicker noise, however, is very sensitive to process changes, so the variations of flicker constant (K_1) may dominate. This is different from the broadband case where the constant was not process dependent. This is the basis for the third rule of thumb.

6.3 Detailed Mathematics for FET Noise

Figure 6.13 illustrates the schematic of the MOSFET and JFET transistor noise models. The fundamental noise relationships for FET transistors are given in Figure 6.14 (Eqs. (6.4) and (6.5)). In this section, we manipulate these equations to show that the rules of thumb also apply to FET transistors. Figure 6.15 shows the thermal noise equation being manipulated for PTAT and zero-TC bias for a FET in strong inversion. Strong inversion refers to the biasing region of the FET. The result of strong inversion is that thermal noise is inversely proportional to the fourth root of I_d. Thermal noise is directly proportional to either the square root or the fourth root of absolute temperature, depending on the bias type. Thus, the FET amplifier in strong inversion is less sensitive to changes in Iq and temperature than the bipolar amplifier.

Figure 6.16 shows manipulation of the thermal noise equation for PTAT and zero-TC bias for a FET in weak inversion. Weak inversion refers to the biasing region of the FET. The result is that thermal noise is inversely proportional to the square root of Id. Thermal noise is directly

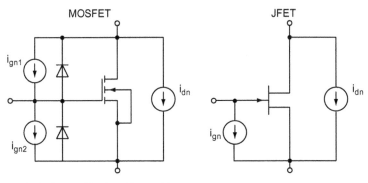

i_{gn} is the combination of i_{gn1} and i_{gn2}.

Figure 6.13: FET transistor noise model

$$i_{gn}^2 = 2 \cdot q \cdot I_G \cdot \Delta f \tag{6.4}$$

where
i_{gn} is gate shot noise current
q is electron charge ($1.6 \cdot 10^{-19}$ coulomb)
I_G is DC gate current (leakage)
Δf is noise bandwidth

$$i_{dn}^2 = 4 \cdot k \cdot T \cdot \left(\frac{2}{3} \cdot g_m \right) \cdot \Delta f + K_3 \cdot \frac{I_D^a}{f^b} \cdot \Delta f \tag{6.5}$$

Thermal noise   Flicker noise

where
i_{dn} is drain noise from shot noise and flicker noise
k is Boltzmann's constant (1.381E-23 joule/°K)
T is temperature in Kelvin
g_m is transconductance of FET
K_3 is process dependent constant for flicker noise
I_D is drain current

Figure 6.14: Fundamental FET noise relationships

proportional to temperature, or the square root of temperature, depending on the bias type. Thus, the FET amplifier in weak inversion has relationships similar to a bipolar bias amplifier for current and temperature.

Figure 6.17 shows the flicker noise equation being manipulated for PTAT and zero-TC bias for a FET in strong inversion. Note that "a" is a constant between 0.5 and 2. Thus, it is possible that flicker noise is proportional to I_d or inversely proportional to some power of I_d

$$e_{nbb} = \sqrt{\frac{2}{3} \cdot \frac{4K \cdot T}{g_m}} \quad \text{Broadband noise for FET}$$

$$g_m = \sqrt{2 \cdot K_S \cdot \left(\frac{W}{L}\right) \cdot I_D} \quad \text{for strong inversion}$$

$$e_{nbb} = \sqrt{\frac{2}{3} \cdot \frac{4K \cdot T}{\sqrt{2 \cdot K_S \cdot \left(\frac{W}{L}\right) \cdot I_D}}}$$

Combine all the constants into K_c

$$e_{nbb} = \sqrt{K_c \frac{T}{\sqrt{I_D}}} \quad \text{zero-TC equation}$$

Note noise for the zero-TC bias is proportional to $\sqrt{T}$
and inversely proportional to $\sqrt[4]{I_D}$

For PTAT substitute, $I_D = I_d \cdot \alpha \cdot T$

$$e_{nbb} = \sqrt{K_c \frac{T}{\sqrt{I_d \cdot \alpha \cdot T}}}$$

Combining all constants into K_d

$$e_{nbb} = K_d \sqrt[4]{\frac{T}{I_d}}$$

Note noise for the PTAT bias is proportional to $\sqrt[4]{T}$
and inversely proportional to $\sqrt[4]{I_D}$

Figure 6.15: FET in strong inversion

depending on the value of "a." For a zero-TC biasing scheme, the value of flicker noise is not dependent on temperature. For a PTAT biasing scheme, the flicker noise will be proportional to the square root of temperature.

Figure 6.18 shows the flicker noise equation being manipulated for PTAT and zero-TC bias for a FET in weak inversion. Note that "a" is a constant between 0.5 and 2. So, for all cases, flicker noise is inversely proportional to some power of I_d. For a zero-TC bias, the flicker noise will be proportional to absolute temperature. For a PTAT bias, the temperature relationship is dependent on the value of a.

$$i_{nbb}^2 = 4k \cdot T \cdot \left(\frac{2}{3} \cdot g_m \right) \text{ Broadband noise}$$

$$g_m = \frac{q \cdot A \cdot I_D}{k \cdot T} \text{ for weak inversion}$$

$$e_{nbb}^2 = \frac{1}{g_m^2} \cdot \left[4k \cdot T \cdot \left(\frac{2}{3} \cdot g_m \right) \right]$$

$$e_{nbb}^2 = \frac{8k \cdot T}{3 \cdot \dfrac{q \cdot A \cdot I_D}{k \cdot T}}$$

$$e_{nbb}^2 = \frac{8k^2 \cdot T^2}{3q \cdot A \cdot I_D}$$

$$e_{nbb} = K_c \frac{T}{\sqrt{I_D}} \text{ for zero-TC}$$

for Ptat, $I_D = \alpha \cdot I_d$

$$e_{nbb}^2 = K_c \frac{T^2}{\alpha \cdot T \cdot I_d}$$

$$e_{nbb} = K_d \sqrt{\frac{T}{I_d}} \text{ for PTAT}$$

Figure 6.16: FET in weak inversion

$$e_{dn}^2 = K_3 \cdot \frac{I_D^a}{g_m^2 f^b} \cdot \Delta f \quad \begin{array}{l} \text{General FET flicker equation} \\ \text{in voltage form} \end{array}$$

$$g_m = \sqrt{2 \cdot K_S \cdot \left(\frac{W}{L} \right) \cdot I_D} \text{ for strong inversion}$$

$$e_{dn} = K_3 \cdot \sqrt{\frac{I_D^{a-1}}{f^b}} \cdot \Delta f \text{ for strong inversion}$$

Figure 6.17: FET flicker noise in strong inversion

$$e_{dn}^2 = K_3 \cdot \frac{I_D^a}{g_m^2 f^b} \cdot \Delta f \quad \text{General FET flicker equation in voltage form}$$

$$g_m = \frac{q \cdot A \cdot I_D}{k \cdot T} \quad \text{for weak inversion}$$

$$e_{dn} = K_3 \cdot \sqrt{\frac{\left(I_D^{a-2}\right) \cdot T^2}{f^b}} \cdot \Delta f \quad \text{for weak inversion}$$

Figure 6.18: FET flicker noise in weak inversion

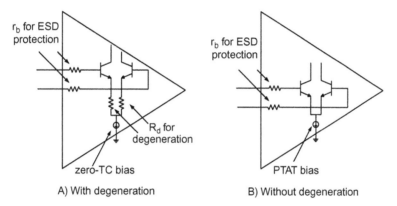

A) With degeneration   B) Without degeneration

Figure 6.19: Simplified op-amp input circuit

6.4 Simplified Physical Connection Inside Amplifier

The equations covered so far are related to a single bipolar or FET transistor. Op-amp designs use many different transistor circuits. The input stage of op-amps always contains a transistor differential amplifier. Figure 6.19 shows two different simplified differential amplifier circuits. Figure 6.19A shows a differential amplifier without degeneration. This circuit would typically use a PTAT current source. Figure 6.19B shows a differential amplifier with degeneration. This circuit would typically use a zero-TC current source. The op-amp noise is affected by biasing schemes, input resistance, transistor size, and process technology.

The input stages shown in Figure 6.19 are simplified. Actual integrated circuit op-amp designs have other transistors that serve many different functions. One common variation of the input stage is the use of two different differential transistor pairs to allow for rail-to-rail input common-mode swing. Figure 6.20 shows a simplified version of this circuit. For part of the common mode, input voltage ranges Q_1 and Q_2 are active. For the other part of the common mode, input voltage ranges Q_3 and Q_4 are active (see Figure 6.20).

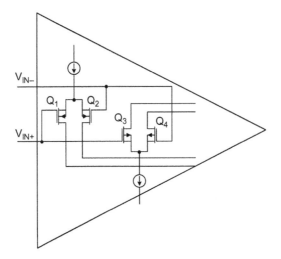

Figure 6.20: Rail-to-rail input stage

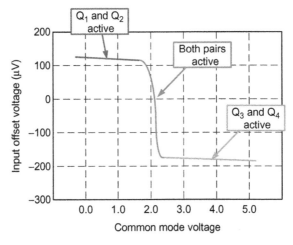

Figure 6.21: Offset voltage vs. common-mode voltage for rail-to-rail input stage

Figure 6.20 shows an example of a typical offset voltage characteristic of the rail-to-rail input topology. Note that the input pair Q_1–Q_2 will have one offset and the input pair Q_3–Q_4 will have a different offset. The same is true for the noise spectral density; i.e., the two different input pairs will have different noise spectral densities (Figures 6.21 and 6.22).

Chapter Summary

- The goal of the chapter is to provide some basic rules of thumb to give board- and system-level designers some insight into the op-amp integrated circuit design.
- The rules of thumb presented in this chapter are common op-amp design practices. Some unique cases may not follow these rules.

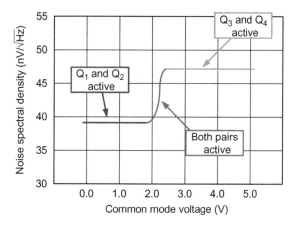

Figure 6.22: Noise spectral density vs. common-mode voltage for rail-to-rail input stage

- Rule 1: Broadband noise voltage is not strongly affected by process variation (i.e., ±10%). Furthermore, broadband current noise may shift as much as two times the typical value (i.e., ±2×).
- Rule 2: Op-amp noise is not strongly affected by temperature changes (i.e., worst case ±33%). Some biasing schemes are less sensitive.
- Rule 3: Flicker noise is highly process dependent (i.e., worst case ±3×).
- Rule 4: Broadband voltage noise is typically lower for larger op-amp I_q.
- Rule 5: FET amplifiers typically have lower current noise than bipolar amplifiers. Furthermore, bipolar amplifiers tend to have lower voltage noise than FET for a given I_q level.
- Rule 5: Bipolar amplifiers tend to have lower voltage noise than FET amplifiers for a given I_q level.
- A rail-to-rail amplifier is an example of a specialized input topology with noise performance that is dependent on common-mode input voltage.

Questions

6.1 What is the worst case broadband noise for the OPA277?

6.2 What is the worst case 1/f noise for the OPA277?

6.3 Which amplifier has lower bias current noise: The OPA227, or the OPA827? Why?

6.4 What design trade-off is made to reduce noise when designing integrated circuit op-amps?

6.5 How does temperature affect amplifier noise?

Further Reading

Gray P.R., Meyer R.G., 1993, Analysis and Design of Analog Integrated Circuits, third ed. Wiley, New York.

CHAPTER 7

Popcorn Noise

This chapter discusses how to measure and identify popcorn noise, the magnitude of popcorn noise as compared to 1/f and broadband noise, and applications that are especially susceptible to popcorn noise.

7.1 Review of 1/f and Broadband Noise

Before looking at popcorn noise, it is useful to review the time domain and statistical representation of broadband and 1/f noise. Both 1/f and broadband noise have Gaussian distributions (see Figures 7.1 and 7.2). Furthermore, these types of noise are consistent and predictable for a given design. Throughout this book, we have learned how to predict noise levels through calculations and simulations. However, these methods cannot be used to measure popcorn noise.

7.2 What Is Popcorn Noise

Popcorn noise is a sudden step or jump in base current on bipolar transistors, or a step in threshold voltage on a FET transistor. It got this name because the sound it makes when played over a speaker resembles the sound of popcorn popping. This noise is also called burst noise and random telegraph signals (RTS). Popcorn noise occurs at low frequency (typically f < 1 kHz). Bursts can happen several times a second, or in some rare cases may take minutes to occur.

Figure 7.3 shows popcorn noise in the time domain and its associated statistical distribution. Note the distinctive jumps in noise level correspond to peaks in the distribution. Clearly, the distribution associated with popcorn noise is not Gaussian. If fact, the distribution shown in this example is three Gaussian curves placed on top of each other (trimodel distribution). This happens because the popcorn noise in this example has three discrete levels. The noise in between bursts is a combination of broadband noise and 1/f noise. Thus, the noise consists of three different Gaussian distributions from 1/f and broadband noise that are shifted to different levels by popcorn noise.

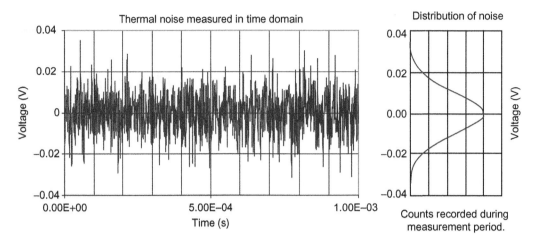

Figure 7.1: Broadband noise—time domain and histogram

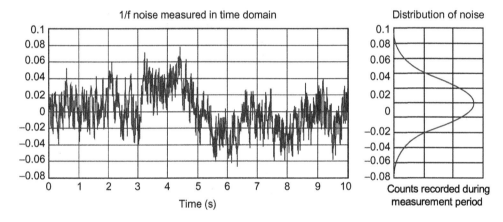

Figure 7.2: 1/f Noise—time domain and histogram

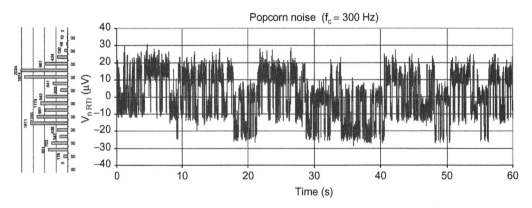

Figure 7.3: Popcorn noise time domain and histogram

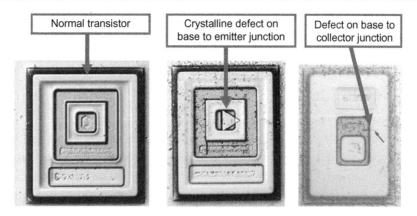

Figure 7.4: Normal transistor vs. transistor with crystalline defect

7.3 What Causes Popcorn Noise?

Popcorn noise is believed to be caused by charge traps or microscopic defects in the semiconductor material. Heavy metal atom contaminates are known to cause popcorn noise. Devices with excessive popcorn noise are often closely examined by experts in the field of failure analysis. Failure analysis searches for microscopic faults that can cause popcorn noise. Figure 7.4 shows how a normal transistor would compare with one that has a crystalline defect.

7.4 How Common Is the Problem?

Popcorn noise is related to problems that occur during semiconductor fabrication. For many modern processes, the occurrence of popcorn noise can be relatively small. Generally, there is a "lot-to-lot" dependency, i.e., some lots will not have popcorn noise while other lots may have a small percentage of devices with popcorn noise. A particularly bad semiconductor lot could have 5% of the devices with popcorn noise. In some cases, it is possible to identify the fabrication issue that caused the popcorn noise.

7.5 Popcorn Noise—Current or Voltage Noise?

On bipolar transistors, popcorn noise shows up as a step change in base current. Therefore, bipolar op-amp popcorn noise typically will show up as bias current noise. For this reason, popcorn noise in bipolar amplifiers may only show up in applications with high source impedance.

On bipolar op-amps with JFET input amplifiers, bias current noise generally is not a problem. In some cases, a bipolar transistor on an internal stage will generate popcorn noise. This popcorn noise shows up as a voltage noise.

In general, MOSFET amplifiers tend to be less prone to popcorn noise. Popcorn noise in MOSFET transistors shows up as a step in the threshold voltage. This would show up as voltage noise in an op-amp.

7.6 Bench and Production Test for Voltage Popcorn Noise

In this chapter, we discuss how to implement a bench test and a production test for popcorn noise. A bench test is a test setup in an engineering laboratory used for testing a small sample of devices. A production test is one that uses automated test equipment to test large quantities of devices. The fundamental difference between the two tests is that the production test needs to have a short test time (typically $t \le 1$ s). The requirement for short test times in production testing is because production test time is very expensive. In many cases, the test cost is comparable to the cost of the semiconductor die.

Figure 7.5 shows the bench setup for measuring voltage popcorn noise of an op-amp (U1). Note that the noninverting input of the amplifier is grounded, so the amplifier's voltage noise and offset is multiplied by the gain. The noise is further amplified by U2. Note that gain of U1 and U2 is set to 100 for both units, i.e., the total gain is $100 \times 100 = 10,000$. This is a typical gain setting for popcorn noise measurements; however, you may need to adjust this for your application.

The low-pass filter (LPF) at the output of U2 limits the bandwidth to 100 Hz. The filter eliminates the higher frequency noise and reveals popcorn noise (or 1/f noise if there is not any popcorn noise). This filter could be adjusted over the range of 10–1000 Hz depending on the application. A 10-Hz low-pass filter has the advantage of attenuating the 60-Hz pickup somewhat. However, it has the disadvantage of obscuring some of the higher frequency bursts. A 1000-Hz low-pass filter will capture higher frequency bursts, but will also begin to include significant broadband noise. The 100-Hz filter is a good compromise between the 10-Hz

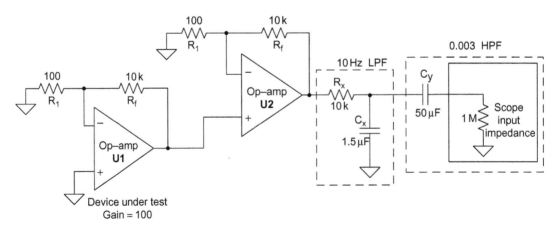

Figure 7.5: Bench test for measuring op-amp voltage popcorn noise

and the 1000-Hz filter. However, you may want to experiment to see what produces the best results for your measurements.

Following U2 is a 0.003-Hz high-pass filter (HPF). The filter is built using a ceramic capacitor and the input impedance of the oscilloscope. Note that several small ceramic capacitors in parallel can be used to build the large ceramic capacitor (e.g., $4 \times 5\,\mu\text{F}$). The high-pass filter is used to eliminate the DC offset. This offset will likely be significantly larger than the noise being measured. The use of this filter allows the noise signal to be measured using the optimal range on the oscilloscope. In this example, the DC output offset is approximately $2\,\text{V}$ and the noise has a 340-mVpp magnitude. The 0.003-Hz HPF removes the 2-V DC component and allows you to observe the 340-mVpp signal on the 200-mV oscilloscope scale.

You can easily estimate the possible output offset by taking the input offset and multiplying it by the total gain. Figure 7.6 shows this calculation. Be careful that the output offset does not drive the amplifier into the power supply rail ($\pm 15\,\text{V}$ for this example). If the output offset approaches the power supply rails, you will need to either reduce the gain or AC couple between U1 and U2. Also note that the filter capacitor C_2 will need to be charged to

Compute the output offset for Figure 7.5

$$V_{out\_offset} = \left[V_{in\_offset1} \cdot (\text{Gain1}) + V_{in\_offset2} \right] \cdot \text{Gain2}$$

$$V_{out\_offset} = [0.2\,\text{mV} \cdot (100) + 0.3\,\text{mV}] \cdot 100$$

$$V_{out\_offset} = 2.03\,\text{V}$$

Compute the delay required to charge the HPF of Figure 7.5

$$\tau = R \cdot C$$

$$\tau = (1\text{M}\Omega)(50\mu\text{F}) = 50\,\text{s}$$

$$5\tau = 5 \cdot (50\text{s}) = 250\,\text{s}$$

for 99.3% settling

Compute the delay required to charge the LPF of Figure 7.5

$$\tau = R \cdot C$$

$$\tau = (10\text{k}\Omega)(1.5\mu\text{F}) = 0.015\,\text{s}$$

$$5\tau = 5 \cdot (0.015\text{s}) = 0.075\,\text{s}$$

for 99.3% settling

Figure 7.6: Calculations associated with op-amp voltage popcorn noise bench test

the output offset voltage when the circuit is initially powered up. This will take a significant amount of time (approximately 5 min). Figure 7.6 also gives the charge time calculation.

Figure 7.7 shows the production setup for measuring voltage popcorn noise of an op-amp (U1). The main difference between the bench setup and the production setup is that digital filters are used in the production test. Digital filters use mathematics to filter the digitized data and consequently they do not have the long charge time associated with analog filters. This keeps the test time short (i.e., low cost). In this example, the tester uses a programmable gain amplifier (PGA) to amplify the noise to a level that is easy to measure. The pedestal digital-to-analog converter (DAC) can be used to cancel the output offset. The final test resources are typical of many production test systems; however, the resources will vary from system to system.

7.7 Bench and Production Test for Current Popcorn Noise

Figure 7.8 shows the bench setup for measuring current popcorn noise of an op-amp (U1). Note that a 1-MΩ resistor is in series with both inputs. The 1-MΩ resistors amplify the

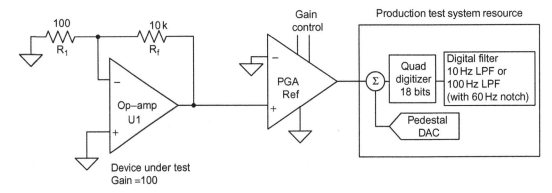

Figure 7.7: Production setup for measuring popcorn voltage noise

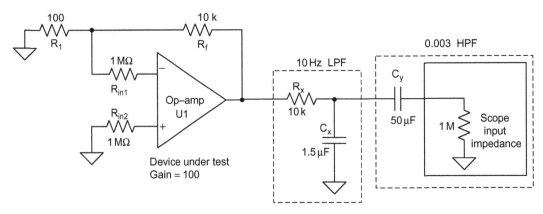

Figure 7.8: Bench test for measuring instrumentation amplifier current popcorn noise

current noise so that it is the dominant noise at the output. Note that this configuration will look for popcorn noise on both inputs. This is important because the noise may be associated with either of the inputs and, consequently, both inputs must be checked. Figure 7.9 illustrates the fact that current noise increases linearly with input resistance and thermal noise increases with the square root of input resistance. Thus, if you increase the input resistance enough, you can always make current noise dominate. Figure 7.10 gives equations to help select an input resistance that will make current noise dominate.

Note that the circuit for measuring current popcorn noise shown in Figure 7.8 does not require the second stage of gain because the input resistors act as a gain to current noise and bias current. The current noise measurement circuit has the same filters that are used in the

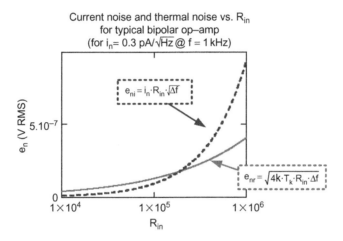

Figure 7.9: Noise increases linearly with input resistance

$$e_{nr} = \sqrt{4k \cdot T_k \cdot R_{in} \cdot \Delta f}$$

$$e_{ni} = i_n \cdot R_{in} \cdot \sqrt{\Delta f}$$

Current noise dominates when its magnitude is at least 3x the thermal noise. Below are the equations solving for R_{in} to make current noise dominate.

Noise from i_n 3× thermal noise

$$i_n \cdot R_{in} \cdot \sqrt{\Delta f} = 3\sqrt{4k \cdot T_k \cdot R_{in} \cdot \Delta f}$$

$$R_{in} = 36 \cdot k \cdot \frac{T_k}{i_n^2}$$

Noise from i_n 5× thermal noise

$$i_n \cdot R_{in} \cdot \sqrt{\Delta f} = 5\sqrt{4k \cdot T_k \cdot R_{in} \cdot \Delta f}$$

$$R_{in} = 100 \cdot k \cdot \frac{T_k}{i_n^2}$$

Figure 7.10: Equations for selecting input resistance

Compute the output offset for Figure 7.8

Typical bipolar values for gain and offset

$I_{b\_offset1} = 0.5$ nA $V_{in\_offset} = 0.3$ mV Gain $= 100$

$V_{out\_offset} = \left[I_{b\_offset1} \cdot (R_{in}) + V_{in\_offset} \right] \cdot$ Gain

$V_{out\_offset} = [0.5$ nA $\cdot (1$ M$\Omega) + 0.3$ mV$] \cdot 100$

$V_{out\_offset} = 0.53$ V

Compute the delay required to charge the HPF of Figure 7.8

$\tau = R \cdot C$

$\tau = (1$M$\Omega)(50\mu$F$) = 50$ s

$5\tau = 5 \cdot (50$s$) = 250$ s

for 99.3% settling

Compute the delay required to charge the LPF of Figure 7.8

$\tau = R \cdot C$

$\tau = (10$k$\Omega)(1.5\mu$F$) = 0.015$ s

$5\tau = 5 \cdot (0.015$s$) = 0.075$ s

for 99.3% settling

Figure 7.11: Calculations pertinent to the filters

voltage noise circuit. The 0.003-Hz high-pass filter eliminates the DC output offset. The DC output offset is generated primarily from the bias current flow through the input resistors. The low-pass filter at the output of U1 limits the bandwidth to 100 Hz. The filter eliminates the higher frequency noise and reveals popcorn noise (or 1/f noise if there is not any popcorn noise). Figure 7.11 gives calculations pertinent to the filters for the current popcorn noise measurement circuit shown in Figure 7.8.

Figure 7.12 shows the production setup for measuring current popcorn noise of an op-amp (U1). The main difference between the bench setup and the production setup is that digital filters are used in the production test. Digital filters use mathematics to filter the digitized data and consequently they do not have the long charge time associated with analog filters. This keeps the test time short (i.e., low cost).

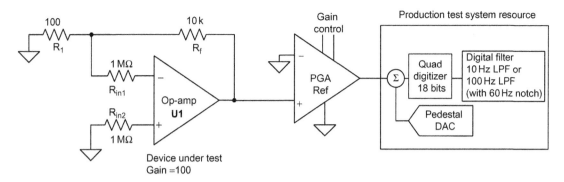

Figure 7.12: Production setup for measuring popcorn current noise

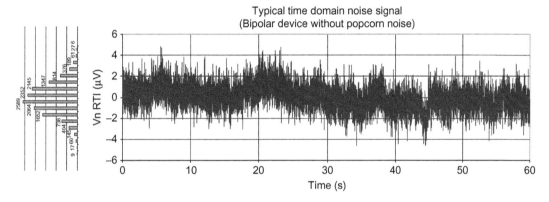

Figure 7.13: Noise for a good unit—1/f noise and broadband noise

7.8 Analyzing the Popcorn Noise Data

The goal of this section is to propose several methods for analyzing low-frequency noise and determining if it contains popcorn noise. The analysis techniques are independent of the circuit configuration used to measure the data. Engineers can often inspect an oscilloscope waveform qualitatively and identify that a signal has popcorn noise. In this section, we propose how to quantitatively identify popcorn noise. Furthermore, we discuss how to set pass/fail limits for popcorn noise and 1/f noise.

Figure 7.13 shows a typical time domain noise signal that does not have popcorn noise. The cut frequency for this signal is 300 Hz, and consequently, the noise is a combination of 1/f noise and broadband noise. The histogram to the left of the noise signal is used to emphasize that the noise voltage is Gaussian. Figure 7.14 is a more detailed view of the Gaussian distribution of the typical noise.

Figure 7.15 shows a typical time domain noise signal that contains popcorn noise. The cut frequency for this signal is 300 Hz. The histogram to the left of the noise signal is used to

Mean = 0.00

Std. dev. = 1.17

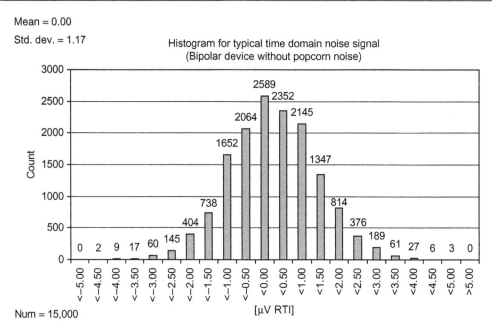

Figure 7.14: Gaussian distribution associated with noise from normal device

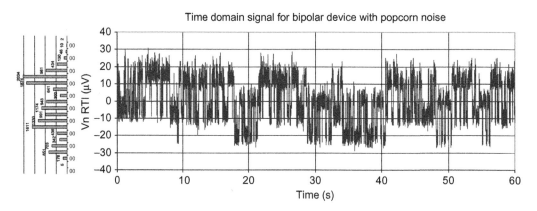

Figure 7.15: Time domain signal of popcorn noise

emphasize that the noise voltage is non-Gaussian. Figure 7.16 is the same waveform shown in Figure 7.15, with circles and arrows used to emphasize the fact that the popcorn noise jumps to discrete modes. For this particular example, there appears to be three discrete levels of noise. These three levels of noise generate three modes in the distribution. A more detailed view of the non-Gaussian distribution of the typical noise is given in Figure 7.17.

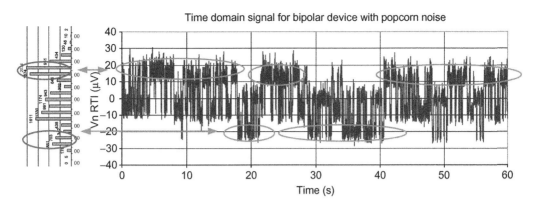

Figure 7.16: Popcorn noise in time domain

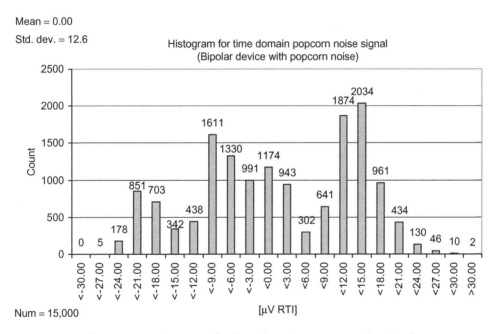

Figure 7.17: Histogram for time domain popcorn noise signal

Therefore, one way to determine if a signal contains popcorn noise is to look for a non-Gaussian distribution. We will not cover the mathematical techniques used to test whether a distribution is Gaussian or non-Gaussian. Instead, we will focus on a technique that looks for the large rapid changes that are associated with the edges of a noise signal. A common way to look for rapid changes in a signal is to take the derivative of the signal. Figure 7.18 shows how the derivative of the popcorn noise signal generates large spikes when the popcorn

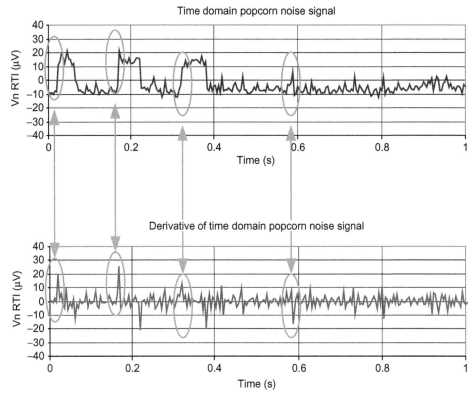

Figure 7.18: Derivative of popcorn noise signal

signal makes a transition. Figure 7.19 shows the derivative of noise from a normal device. The noise in Figure 7.19 is only broadband and flicker popcorn noise, i.e., no popcorn noise. Note that taking the derivative of broadband and flicker noise does not generate the large spikes.

Figure 7.20 compares the derivative histogram of the popcorn noise to the derivative histogram of the noise from a normal device. The histogram from popcorn noise has a large number of counts in the outlying bins. These outliers correspond to the spikes in the derivative. Note that the histogram from the normal device does not have significant number of outliers. For this example, we look for outliers in the distribution at $\pm 4\sigma$. The statistical probability of measuring noise outside of ± 4 standard deviations is 0.007%. The example histogram shown contains 15,000 samples, so we should expect no more than one sample (i.e., $15,000 \times 0.007\% = 1.05$) outside these limits. Thus, excessive bins outside the $\pm 4\sigma$ limits are likely popcorn noise. The limits for this test should be adjusted based on the number of samples in the histogram.

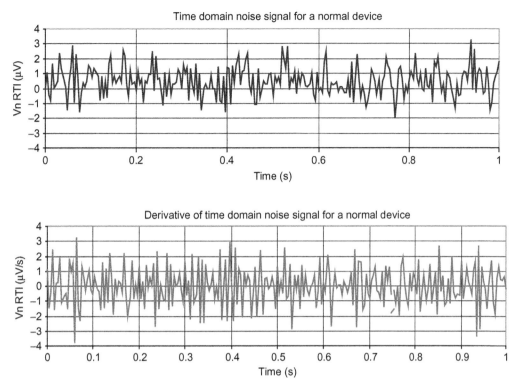

Figure 7.19: Derivative of noise from a normal device

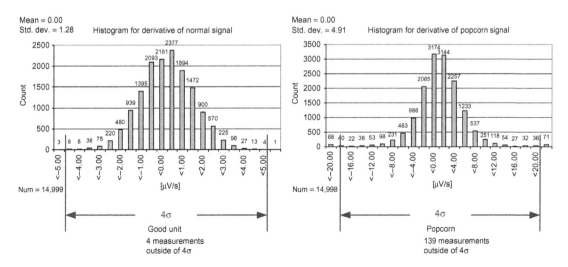

Figure 7.20: Distribution of the noise derivative for normal and popcorn device

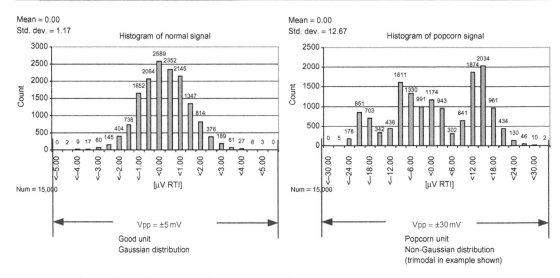

Figure 7.21: Comparing the peak-to-peak noise of a normal vs. popcorn device

Another way to search for devices with popcorn noise is to compare the measured peak-to-peak noise to the expected peak-to-peak noise. Figure 7.21 compares the distribution of a device with popcorn noise to a normal device. Note that the popcorn noise peak-to-peak is six times the normal unit. Note that the scale is adjusted so that the non-Gaussian nature of the popcorn noise is emphasized. Keep in mind that abnormally large low-frequency noise is a strong indication of popcorn noise, but it does not necessarily prove the existence of popcorn noise. However, devices that have abnormally high noise levels are problematic, regardless of whether or not the noise is popcorn noise.

7.9 Setting Limits to a Popcorn Noise Test

This chapter proposes two methods for screening out popcorn noise. The first method involves taking the derivative of the noise signal and searching for outliers in the distribution. The limit proposed for this test is ±4 standard deviations. Thus, if any point in the derivative exceeds ±4 standard deviations, the device is considered to be a failure.

The second method proposed for popcorn nose screening is to look at the peak-to-peak noise. The limit for this test should be set using the worst-case noise rules from Chapter 6 of this book. Figure 7.22 summarizes these rules graphically. The worst-case rules of thumb describe how the spectral density curve can change with process variations. Using the methods described in Chapter 3 and the worst-case spectral density curve, you can estimate the

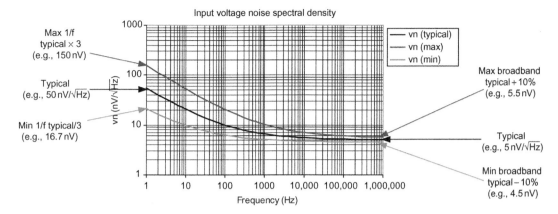

Figure 7.22: Worst-case rule of thumb for noise

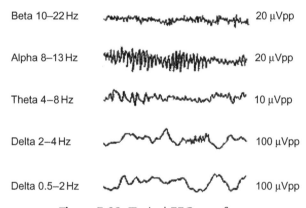

Figure 7.23: Typical EEG waveforms

worst-case expected noise. The worst-case expected noise is the maximum noise expected with a normal device. Devices with popcorn noise typically exhibit noise greater than the worst-case limit. Thus, peak-to-peak noise limits should be set to the worst-case estimate. Devices failing these limits will either have popcorn noise or have excessively high flicker; in either case, they should be considered failures.

7.10 When Is Popcorn Noise a Concern?

Popcorn noise is a concern for low-frequency applications (fc < 1 kHz) of slow-moving signals. For example, the frequency range and waveforms in medical encephalogram (EEG, brain measurements) would be difficult to discern from popcorn noise. Figure 7.23 shows

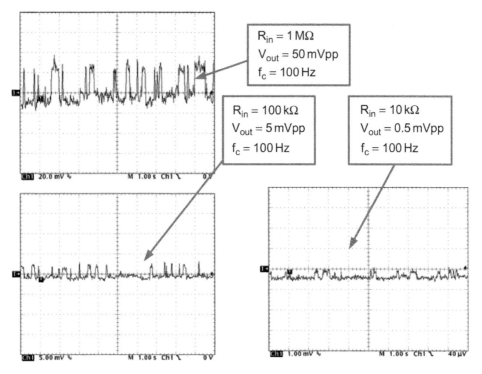

Figure 7.24: Current popcorn noise affected by source impedance

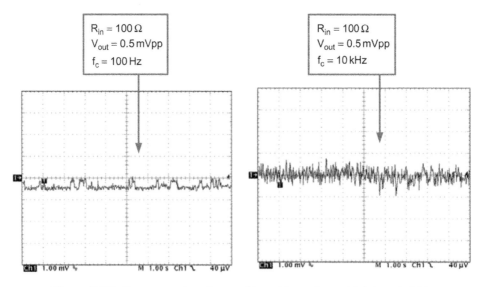

Figure 7.25: Popcorn noise obscured by white noise at high bandwidth

typical EEG waveforms. Seismic measurements are also slow-moving DC signals that can be difficult to discern from popcorn noise. In audio applications, popcorn noise is considered to be a particularly unpleasant noise.

Popcorn noise often shows up as a current noise, and so high-source impedance applications may be more susceptible to popcorn noise. Figure 7.24 shows how the magnitude of popcorn is affected by input impedance. Keep in mind, however, that in some cases internal current noise is converted to voltage noise inside the device.

In some cases, popcorn noise may be obscured by broadband noise. Figure 7.25 shows the same device for two different bandwidths. Note that both waveforms in Figure 7.25 contain popcorn noise, but the popcorn in the wide bandwidth case is obscured by white noise.

Chapter Summary

- Popcorn noise is of low frequency (0–1 kHz).
- Popcorn noise is a step change in the noise level.
- 1/f Noise and broadband noise have a Gaussian distribution. Popcorn noise has a bimodal or multimodel distribution.
- 1/f Noise and bipolar noise can be predicted using op-amp specifications, but popcorn noise cannot be predicted.
- Popcorn noise is caused by semiconductor defects or processing issues.
- A device with popcorn noise is considered to be a failure. 1/f noise and broadband noise, however, are normal and present in all op-amps.
- Popcorn noise is process related. Thus, some wafer lots will not have any popcorn noise and others will have some percentage of devices with popcorn noise.
- Popcorn noise is most frequently seen in bipolar processes.
- Popcorn noise is frequently a bias current noise. Reducing input impedance can often minimize noise.
- Popcorn noise is an important concern in low-frequency, high-gain applications (e.g., EEG, seismic).

Questions

7.1   Assume an op-amp has $1\,\text{fA}/\sqrt{\text{Hz}}$ noise. What value of input resistance would be required to measure current noise? Keep in mind that the current noise should be at least three times greater than the resistors' thermal noise.

7.2   (a) For the circuit below, what is the expected peak-to-peak output noise? (b) What is the maximum peak-to-peak noise? (c) What peak-to-peak noise level would indicate that the device has popcorn noise? (d) What are some other ways to prove that the device has popcorn noise?

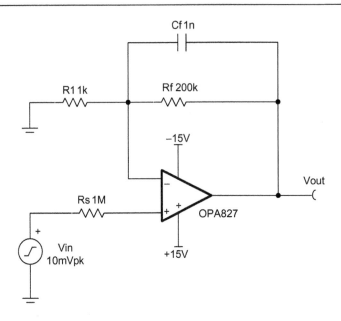

Further Reading

Gray, P.R., Meyer, R.G., 1993, Analysis and Design of Analog Integrated Circuits, third ed. Wiley, New York.

CHAPTER 8

1/f Noise and Zero-Drift Amplifiers

This chapter focuses on errors in low-frequency applications. Input offset voltage drift and 1/f noise are studied in detail. Standard topologies are compared and contrasted with the zero-drift amplifier topology, which has low offset drift and no 1/f noise.

8.1 Zero-Drift Amplifiers

Zero-drift amplifiers are operational amplifiers (op-amps) that periodically self-calibrate offset, offset drift, and low-frequency noise errors. The calibration period for different zero-drift amplifiers ranges from 10 kHz to 100 kHz. A digital circuit inside of the zero-drift amplifier controls the calibration; however, the amplifier acts as a normal linear op-amp from the board- and system-level view. There are two different types of zero-drift amplifiers: auto-zero and chopper stabilized.

The auto-zero amplifier consists of a continuous linear amplifier and a nulling amplifier. During calibration, the nulling amplifier samples the offset voltage and stores it on a sampling capacitor. The linear amplifier's offset is canceled by the offset stored on the sampling capacitor. Since the calibration is done at a relatively high rate (e.g., 50 kHz), the offset drift and low-frequency noise are also canceled. Figure 8.1 shows a simplified block diagram of

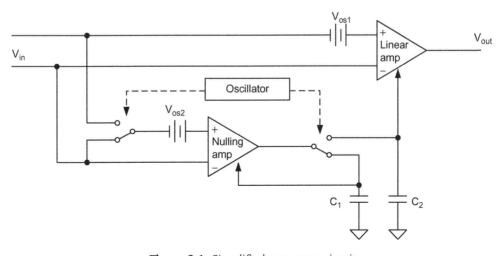

Figure 8.1: Simplified auto-zero circuit

an auto-zero circuit. Although this is an oversimplification of the auto-zero correction, it is adequate from a functional board- and system-level designer view. For a more detailed description of how this technique works refer to the first reference listed under the "Further Reading" section of this chapter.

In the chopper-stabilized circuit, the input and output are synchronously inverted. Thus, the offset is inverted at every other chopping cycle. This converts the offset from a constant DC value to an AC signal with an average of zero. Filtering reduces the amplitude of the AC signal created by the offset chopping (see Figure 8.2). Modern Texas Instruments chopper amplifiers use a patented switched capacitor notch filter to eliminate the chopper signal. For a more detailed description of how the chopping technique works refer to the second reference listed under the "Further Reading" section of this chapter.

Both the auto-zero and chopper-stabilized amplifiers share some common characteristics and are categorized as zero-drift amplifiers. The key characteristics are low-voltage offset and low-voltage-offset drift. Although bias current and bias current drift are not calibrated during the self-calibration, they are typically low because the amplifiers are MOSFET amplifiers. Table 8.1 lists offset and offset drift for some common zero-drift amplifiers.

Another important characteristic of zero-drift amplifiers is that they have practically no 1/f noise. Low-frequency noise can be thought of as variations in offset voltage with time. The self-calibration eliminates low-frequency noise in the same way it eliminates offset drift. Figure 8.3 shows a noise waveform sampled and corrected over time.

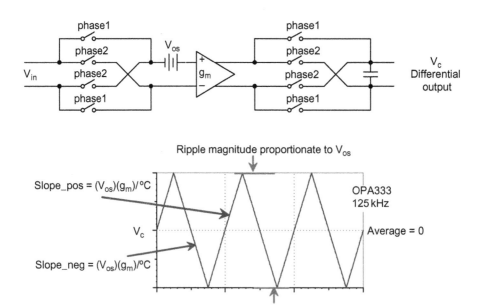

Figure 8.2: Chopper-stabilized amplifier and AC signal created from chopping

Table 8.1: Offset and Offset Drift for Common Op-amp

Op-amp	Offset (μV)	Offset Drift (μV/C)	Voltage Noise (nV/$\sqrt{Hz}$)
OPA333	10	0.05	55
OPA335	5	0.05	55
OPA378	50	0.25	20

Offset calibration occurs at an instant in time.

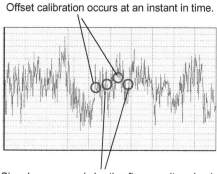

Signals processed shortly after are altered only by high-frequency noise.

Figure 8.3: Low-frequency noise eliminated by zero-drift amplifier

8.2 Zero-Drift Amplifier Spectral Density Curve

The spectral density curve for zero-drift amplifiers does not have a 1/f region. In some cases, the calibration process creates signals at the calibration frequency and harmonics. These signals show up as spikes on the spectral density curve. Figure 8.4 shows the spectral density curve for a typical auto-zero amplifier.

For most applications, it is a good idea to avoid the region with the calibration signal feedthrough. This can be done with an external filter. In many cases, the op-amp gain bandwidth automatically attenuates the calibration feedthrough. In Figure 8.5, the 3-dB bandwidth is limited to 2 kHz by the gain bandwidth limitation of the amplifier; i.e., Gain_Bandwidth/Gain = 2 MHz/1001 = 2 kHz. In this configuration, the calibration signal is significantly attenuated. Figure 8.6 shows the noise calculation for the gain of 1001. Note that in this configuration, the calculated total noise closely matches the measured total noise. For this configuration, the noise is flat throughout the entire frequency of operation; i.e., 0–2 kHz. The mathematics used in this calculation is covered in Chapter 2 of this book.

Figure 8.7 shows the noise in gain of 101. In this case, the 3-dB bandwidth is limited to 19.8 kHz by the gain bandwidth limitation of the amplifier; i.e., Gain_Bandwidth/

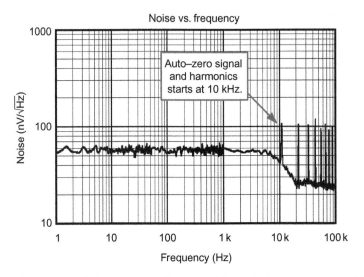

Figure 8.4: Spectral density curve for OPA335 (typical auto-zero amplifier)

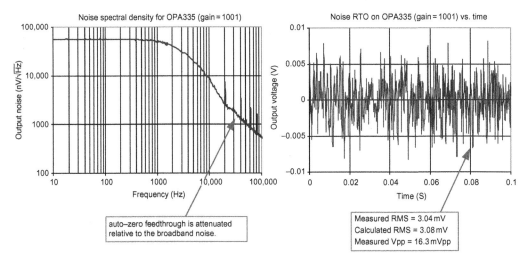

Figure 8.5: Measured output noise for OPA335 in gain of 1001

$$f_H = \frac{Gain\_Bandwidth}{Gain} = \frac{2\,MHz}{1001} = 1.998\,kHz$$

$$BW_n = f_H \cdot K_n = 1.998\,kHz \cdot 1.57 = 3.137\,kHz$$

$$V_{n\_out} = e_{BB} \cdot \sqrt{BW_N} \cdot Noise\_Gain = 55\,nV \cdot \sqrt{3.137\,kHz} \cdot (1001) = 3.08\,mV$$

Figure 8.6: Total noise calculation for OPA335 in gain of 1001

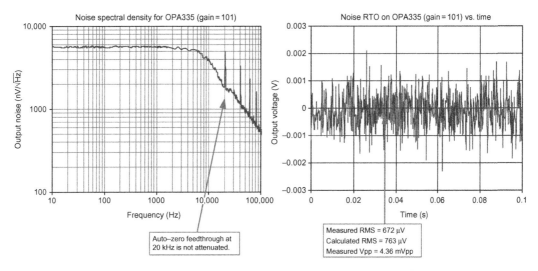

Figure 8.7: Measured output noise for OPA335 in gain of 101

Gain $= 2\,\text{MHz}/101 = 19.8\,\text{kHz}$. Since the bandwidth is $19.8\,\text{kHz}$, the calibration signal at $10\,\text{kHz}$ is not attenuated. Also note that the calculation for total noise is complicated by the fact that the spectral density curve drops from $55\,\text{nV}/\sqrt{\text{Hz}}$ to $25\,\text{nV}/\sqrt{\text{Hz}}$ at $10\,\text{kHz}$. The calibration signals are included in the noise signal, and cannot be easily accounted for in the calculation. The total noise is calculated in Figure 8.8.

Figure 8.9 shows the noise in gain of 11. In this case, the 3-dB bandwidth is limited to $182\,\text{kHz}$ by the gain bandwidth limitation of the amplifier so the calibration is not attenuated. Figure 8.10 compares the three gains considered in this chapter.

Modern Texas Instruments chopper amplifiers use filtering techniques that minimize the calibration signal feedthrough. Another trick used in modern op-amp designs is to move the calibration frequency to a higher frequency to increase the usable bandwidth. Figure 8.11 shows the spectral density curve for the OPA333 chopper-stabilized amplifier. The chopper frequency for this amplifier is approximately $125\,\text{kHz}$. The gain bandwidth for the OPA333 is $350\,\text{kHz}$. For most gain settings, the chopper frequency is outside of the bandwidth of the amplifier. For example, the bandwidth would be limited to $35\,\text{kHz}$ for a gain of 10, and consequently, the chopper signal would be substantially attenuated.

8.3 Low-Frequency Noise

The spectral density curve for 1/f noise increases as frequency is decreased. In fact, noise will increase toward infinity as we approach zero frequency. This fact often leads designers to believe that the total noise should be infinite at DC, because DC is considered to be zero

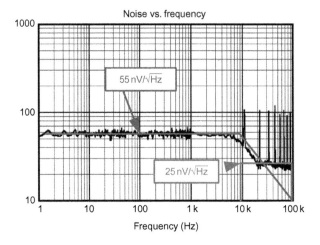

$$BW_{n1} = f_H \cdot K_n = 10\,kHz \cdot 1.57 = 15.7\,kHz$$

$$V_{n\_out1} = e_{BB} \cdot \sqrt{BW_{n1}} \cdot Noise\_Gain = 55 \cdot nV \cdot \sqrt{15.7\,kHz} \cdot (101) = 696\,\mu V$$

$$f_H = \frac{Gain\_Bandwidth}{Gain} = \frac{2\,MHz}{101} = 19.82\,kHz$$

$$BW_{n2} = f_H \cdot K_n = (19.82\,kHz - 10\,kHz)1.57 \cdot = 15.7\,kHz$$

$$V_{n\_out2} = e_{BB} \cdot \sqrt{BW_{n2}} \cdot Noise\_Gain = 25\,nV \cdot \sqrt{15.3\,kHz \cdot 1.57}(101) = 313\,\mu V$$

$$V_{n\_total} = \sqrt{V_{n1}^2 + V_{n2}^2} = \sqrt{(696\,\mu V)^2 + (313\,\mu V)^2} = 763\,\mu V$$

Figure 8.8: Total noise calculation for OPA335 in gain of 101

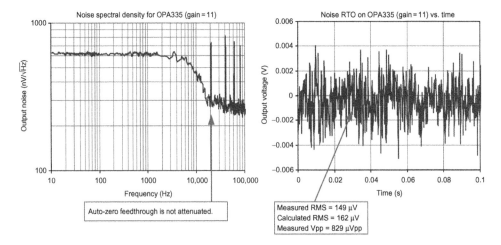

Figure 8.9: Measured output noise for OPA335 in gain of 11

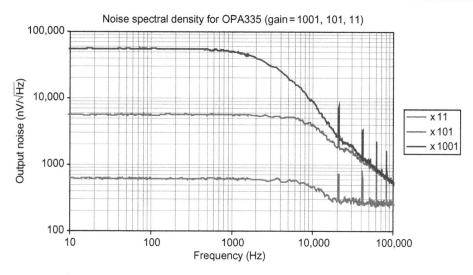

Figure 8.10: Comparing output spectral density for different gains

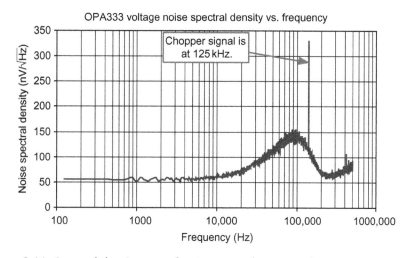

Figure 8.11: Spectral density curve for OPA333 (chopper with internal notch filter)

frequency. The best way to understand why 1/f noise does not translate to infinite noise for practical circuits is to convert frequency to time (i.e., Time = 1/Frequency).

Table 8.2 shows flicker noise (1/f noise) calculations for the OPA336 at different low-cut frequencies. The lower cut frequency is set by the time period that the signal is observed over. Typically, noise calculations use 0.1 Hz that is used as a lower cut frequency. This corresponds to an observation period of 10 s. The same calculation can be made for any time period. Note that a frequency of 0 Hz corresponds to infinite time and, consequently, is not practically achievable. Note that extremely low frequencies correspond to years of time. Figure 8.12 shows how noise increases for longer observation periods. Figure 8.12 is the graphical representation of Table 8.2.

The upper cut frequency used in these calculations is $10\,\text{Hz}$. This is the typical upper cut frequency used in flicker noise measurements (i.e., $0.1\,\text{Hz} < f < 10\,\text{Hz}$). The noise calculation computes the noise from f_L to f_H, assuming a "brick wall filter" at f_L and f_H.

Table 8.2: Flicker Noise Calculations for the OPA336

f_H	f_L	$1/f_L$ (s)	$1/f_L$ (days)	$1/f_L$ (years)	Noise Calculation	Noise (nV)
10	1	1	1.1×10^{-5}	3.1×10^{-8}	$200\text{nV} \cdot \sqrt{\ln\left(\dfrac{10\text{Hz}}{1\text{Hz}}\right)}$	303
10	0.1	10	1.1×10^{-4}	3.1×10^{-7}	$200\text{nV} \cdot \sqrt{\ln\left(\dfrac{10\text{Hz}}{0.1\text{Hz}}\right)}$	429
10	1×10^{-6}	1×10^{6}	11	0.032	$200\text{nV} \cdot \sqrt{\ln\left(\dfrac{10\text{Hz}}{1\mu\text{Hz}}\right)}$	808
10	1×10^{-9}	1×10^{9}	1.1×10^{4}	32	$200\text{nV} \cdot \sqrt{\ln\left(\dfrac{10\text{Hz}}{1\text{nHz}}\right)}$	960

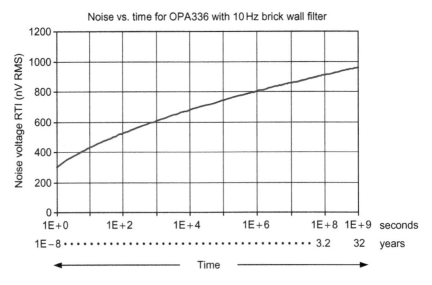

Figure 8.12: Noise for OPA336 vs. time ($F_H = 10\,\text{Hz}$)

This is the way 0.1 Hz to 10 Hz noise is normally specified. The term "brick wall filter" means that the noise drops abruptly to zero outside the specified bandwidth. If a real-world filter is used, the noise decreases gradually with the filter (i.e., 20 dB/decade for a first-order filter). This topic is covered in detail in Chapter 2 of this book.

The total noise from flicker (1/f noise) is equal over each decade change in frequency. For example, the total noise in the interval "0.1 Hz, 10 Hz" is the same as in the interval "0.01 H, 0.1 Hz"; this is shown mathematically in Figure 8.13 using the formula developed in Part 1. This fact is often confusing to engineers because the area appears to be significantly larger in regions with higher flicker noise. However, keep in mind that spectral density curves are usually shown with a logarithmic axis. When you look at the area of two different decade-wide intervals on a logarithmic axis, the intervals do not look equivalent. If you change to a linear axis, you see that as 1/f noise gets larger, the width of the interval gets smaller. Figure 8.14 shows the

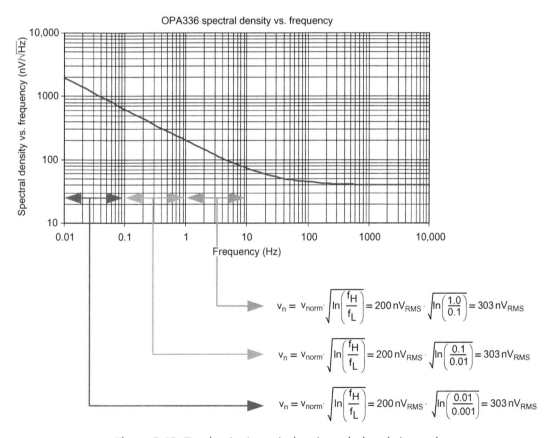

$$v_n = v_{norm} \cdot \sqrt{\ln\left(\frac{f_H}{f_L}\right)} = 200\,nV_{RMS} \cdot \sqrt{\ln\left(\frac{1.0}{0.1}\right)} = 303\,nV_{RMS}$$

$$v_n = v_{norm} \cdot \sqrt{\ln\left(\frac{f_H}{f_L}\right)} = 200\,nV_{RMS} \cdot \sqrt{\ln\left(\frac{0.1}{0.01}\right)} = 303\,nV_{RMS}$$

$$v_n = v_{norm} \cdot \sqrt{\ln\left(\frac{f_H}{f_L}\right)} = 200\,nV_{RMS} \cdot \sqrt{\ln\left(\frac{0.01}{0.001}\right)} = 303\,nV_{RMS}$$

Figure 8.13: Total noise is equivalent in each decade interval

power spectral density curve on a linear axis to illustrate the equivalent area of two decade-wide intervals.

The total noise for amplifiers with 1/f noise increases for longer observation time periods. The waveform in Figure 8.15 shows the noise on the OPA336 over a 100,000-s interval (10 µHz). The upper cut frequency for this signal is 10 Hz. Thus, the noise bandwidth is 10 µHz to 10 Hz. The total RMS noise over the entire interval is 0.74 µV. If you choose any subinterval of time, the total RMS noise will be smaller. In this example, a 10-s subinterval is shown to have a total noise of 0.43-µV RMS. The subinterval in this example was taken from the first 10 s, but any 10-s interval will have the same total RMS noise. Remember that a smaller time period corresponds to a larger lower cut frequency, and less area under the 1/f curve.

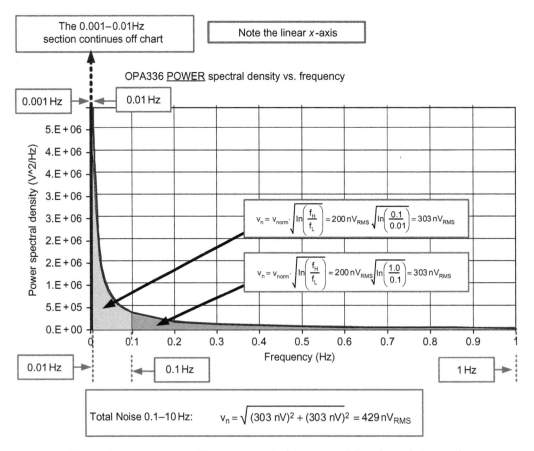

Figure 8.14: Linear axis illustrates equivalent area of decade-wide intervals

Zero-drift amplifiers do not contain flicker noise, so total noise is computed using the same method as broadband noise. Since noise spectral density is flat, it is possible to integrate noise down to 0 Hz. It is not possible to integrate down to 0 Hz with flicker noise because spectral density is infinite at 0 Hz. The total noise in each decade-wide subinterval is equal for flicker noise. With broadband noise, the total noise dramatically decreases for lower subintervals.

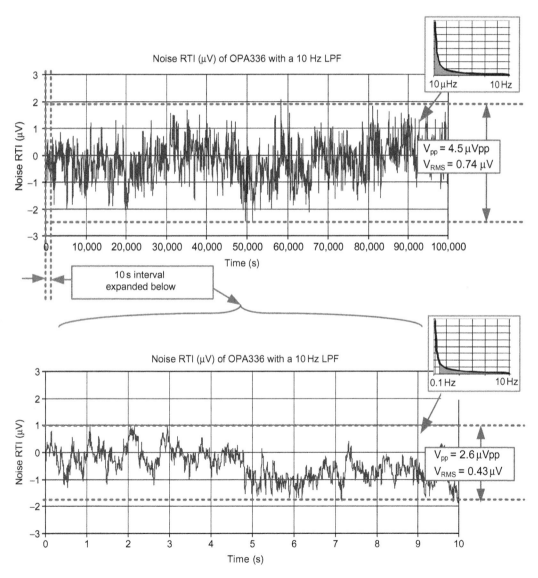

Figure 8.15: OPA336 noise over a long time interval

Looking at the power spectral density curve on a linear axis helps illustrate how total noise decreases for lower subintervals (see Figure 8.16).

Table 8.3 shows noise calculations for the OPA333 at different lower cut frequencies. The lower cut frequency is set by the observation period. Note that there is a very little change in total noise with time. Because the spectral density is flat, the lower frequency subintervals have very little area (total noise). This is an advantage of the zero-drift topology compared to devices with flicker noise. Figure 8.17 shows that noise from the zero-drift amplifier is unchanged out to extremely long times.

The total noise for zero-drift amplifiers remains constant for different observation periods. The waveform in Figure 8.18 illustrates the OPA333 over a 100,000-s interval ($10\,\mu Hz$). The upper cut frequency for this signal is $10\,Hz$. Thus, the noise bandwidth is $10\,\mu Hz$ to $10\,Hz$. The total RMS noise over the entire interval is $0.173\,\mu V$. If you choose any subinterval of time, the total

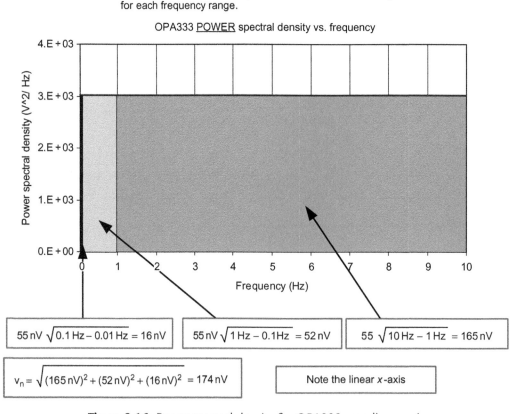

Figure 8.16: Power spectral density for OPA333 on a linear axis

Table 8.3: OPA333 Noise Over a Long Time Interval

f_H	f_L	$1/f_L$ (s)	$1/f_L$ (days)	$1/f_L$ (years)	Noise Calculation	Noise (nV)
10	1	1	1.1×10^{-5}	3.1×10^{-8}	$55nV\sqrt{10Hz - 1Hz}$	165
10	0.1	10	1.1×10^{-4}	3.1×10^{-7}	$55nV\sqrt{10Hz - 0.1Hz}$	173
10	1×10^{-6}	1×10^{6}	11	0.032	$55nV\sqrt{10Hz - 1\mu Hz}$	174
10	1×10^{-9}	1×10^{9}	1.1×10^{4}	32	$55nV\sqrt{10Hz - 1nHz}$	174

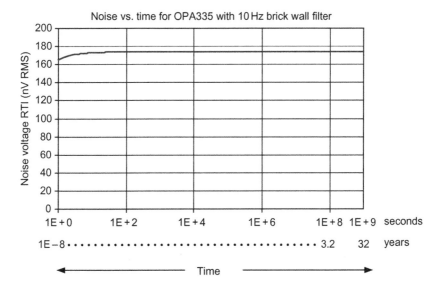

Figure 8.17: OPA333 noise over a long time interval

RMS noise will be the same. In this example, a 10-s subinterval is shown to have a total noise of 0.173-μV RMS. The subinterval in this example was taken from the first 10s, but any 10-s interval will have the same total RMS noise. Remember that the total noise for the two cases is very close because the area under the power spectral density curve is nearly the same.

8.4 Measuring Low-Frequency Noise

Most data sheets specify low-frequency noise as noise over the interval 0.1–10 Hz. An active filter circuit for measuring this noise is covered in Chapter 5. This chapter discusses low-frequency noise with extremely low cut frequencies (e.g., 10 μHz). Measuring extremely low frequencies makes it impractical to AC-couple the signal because the component values are outside of usable ranges. Figure 8.19 shows a DC-coupled circuit that can be used to measure

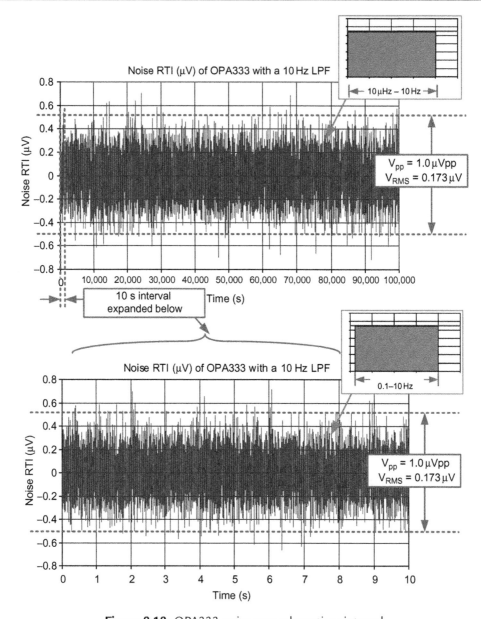

Figure 8.18: OPA333 noise over a long time interval

low-frequency noise. Depending on the noise level of the amplifier under test, the gain of this circuit should be adjusted.

One problem with this circuit is that it also amplifies the DC offset of the amplifiers. In the example in Figure 8.19, noise is gained to a level that can be easily read on an oscilloscope

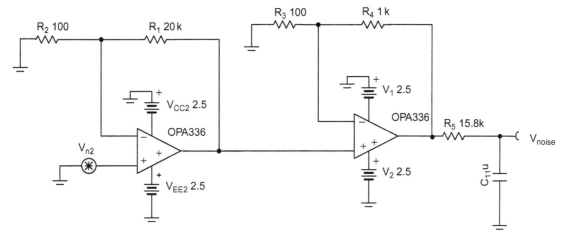

Figure 8.19: Circuit for measuring noise (DC to 10 Hz)

Total output flicker noise from 0.1–10 Hz

$$V_{noise\_RMS} = V_n \cdot 1\,Hz \sqrt{\ln\left(\frac{f_H}{f_L}\right)} \cdot Gain$$

$$V_{noise\_RMS} = 200\,nV \cdot \sqrt{\ln\left(\frac{10}{0.1}\right)} \cdot 201 \cdot 11 = 0.949\,\mu V_{RMS}$$

$$Noise\_pp = 6\,V_{noise\_RMS} = (0.949\,\mu V_{RMS}) \cdot 6 = 5.69\,mVpp$$

Output offset voltage

$$V_{out\_offset} = V_{off} \cdot Gain1 \cdot Gain2 = (500\,\mu V) \cdot (201) \cdot 11 = 1.11\,V$$

Figure 8.20: Noise and offset calculation from Figure 8.22

(i.e., 5.69 mVpp, see Figure 8.20). Make sure when choosing gain that gain of the first stage is at least 10; this makes the noise from the first stage dominate. In this example, the offset is gained to $2\,V_{DC}$; this makes it impossible to measure the noise on the appropriate range. The circuit shown in Figure 8.21 can be used to correct this offset.

One important consideration when making low-frequency noise measurements is to ensure that offset temperature drift does not affect the results. Offset voltage temperature drift looks

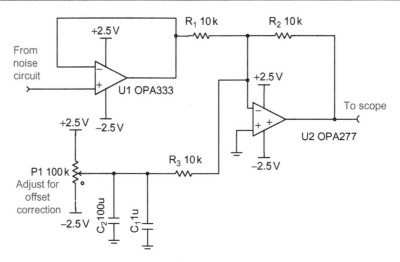

Figure 8.21: Circuit for canceling output offset

very similar to 1/f noise. In fact, it is impossible to distinguish between the two error sources by analyzing the output signal. Figure 8.22 shows the noise plus offset drift on the OPA336 over 10,000 s. No temperature control was used during this measurement (the ambient lab temperature varied 2 °C). This small fluctuation in room temperature significantly affected the output signal. Figure 8.23 compares the results of the circuit with no temperature control to the same circuit under precision control. The precision temperature control is achieved using a thermal bath. The thermal bath used in this measurement is a chamber filled with an inert, fluorinated fluid controlled to 0.01 °C. The typical expected drift for this example is $1.5 \mu V \times 1.6 = 2.4 \mu V$. In this case, the drift appears to be about $5 \mu V$; i.e., the mean value of the signal shifts from $0 \mu V$ to $-5 \mu V$.

In the case of zero-drift amplifiers, the effect of offset temperature drift is greatly reduced. Figure 8.24 shows noise measurements made with no temperature control (the ambient lab temperature varied 2 °C). Figure 8.25 compares the measurement with no temperature control to the same measurement made in a thermal bath. There is no noticeable difference between the circuit with and without temperature control. The ultralow temperature drift of the zero-drift amplifier is a key benefit of this topology.

Another important consideration for noise measurement is to measure the RMS noise (i.e., standard deviation) rather than peak-to-peak noise. Often, people read noise as a peak-to-peak reading on the oscilloscope. This is good as an approximation, but is not adequate for precision results. The problem is that the number of samples can greatly affect the peak-to-peak reading. Remember that peak-to-peak can be estimated by multiplying the standard

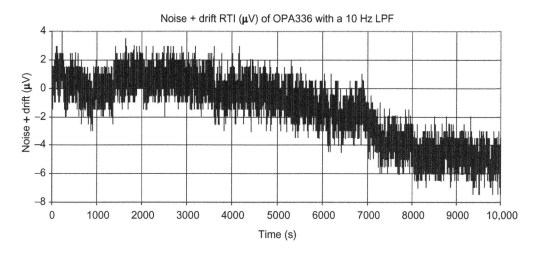

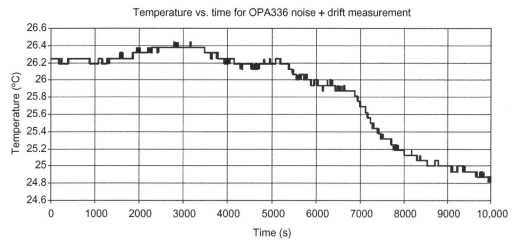

Figure 8.22: OPA336 noise plus temperature drift measured in ambient lab temperature

deviation by six. Mathematically, this means that there is a 99.7% chance that the noise is bounded by peak-to-peak estimate. So, 0.3% of the noise is outside this range. If the number of samples is increased, you will see more occurrences outside the six-sigma peak-to-peak estimate. The RMS noise, however, remains relatively constant regardless of the number of samples. Figure 8.26 illustrates the same signal captured on a digitizing oscilloscope with 25,000 samples and 2500 samples. The peak-to-peak measurement is significantly larger for the chart with more samples. Chapter 1 gives more detail on this topic.

One final note regarding RMS noise measurements is that you should be careful *not* to include DC into the noise measurement (i.e., average = 0). The mathematics of noise

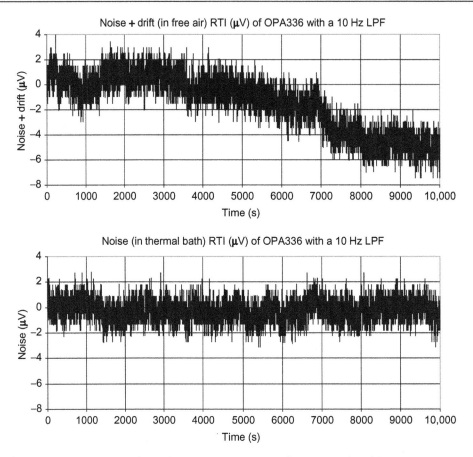

Figure 8.23: OPA336 noise tight temperature control vs. normal ambient temperature

analysis assumes that the average value is zero. The easiest way to eliminate the average is to find the standard deviation of the signal rather than the RMS. The standard deviation is mathematically defined to be the RMS with zero average (see Figure 8.27). Most digitizing oscilloscopes have the ability to save the results into a spreadsheet format. In the spreadsheet, you can use spreadsheet mathematics to compute the standard deviation, e.g., " = STDEV (..RANGE..)" in Microsoft Excel®. Some digitizing oscilloscopes provide an RMS mathematical operator. However, in general, you will get more accurate results by importing the data into a spreadsheet and perform the standard deviation. This is because the signal typically has a slight DC offset even when AC coupled. Of course, different instruments have different idiosyncrasies, so experiment with your equipment to learn its limitations.

Chapter Summary

• Zero-drift amplifiers use an internal digital calibration circuit to correct offset voltage and offset drift.

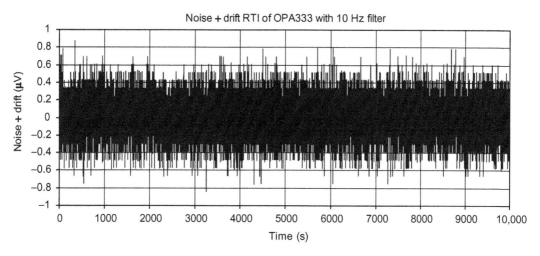

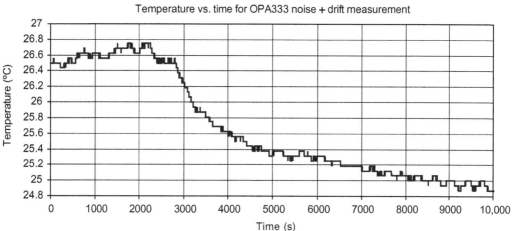

Figure 8.24: OPA333 noise plus temperature drift measured in ambient lab temperature

- Two common zero-drift topologies are auto-zero and chopper stabilized.
    - Auto-zero amplifiers sample and hold the offset and use the sampled value to correct offset error.
    - Chopper-stabilized amplifiers synchronously flip the polarity of the input and output. This effectively passes signals through the amplifier unaffected, but converts the offset to an AC signal with an average of zero.
- Zero-drift amplifiers eliminate 1/f noise during the calibration process.
- The calibration in a zero-drift amplifier occurs at a specific frequency.
    - Noise at the calibration frequency can be seen in the spectral density curve as a spike.
    - In some cases, multiple spikes or harmonics of the calibration signal will be present.

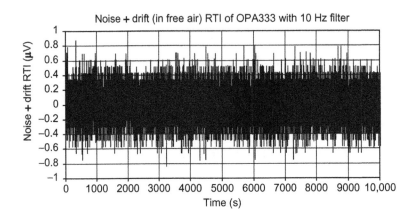

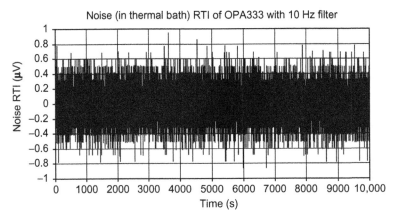

Figure 8.25: OPA333 noise tight temperature control vs. normal ambient temperature

- In most modern amplifiers, the calibration signal is designed to be at a frequency near the bandwidth of the amplifier. So for most gains, the calibration signal will be substantially attenuated.
- The peak-to-peak noise for amplifiers with 1/f noise will continue to increase the longer it is monitored. Remember that noise goes to infinity as frequency goes to zero for amplifiers with 1/f noise.
- The peak-to-peak noise for amplifiers without 1/f noise (e.g., zero-drift amplifiers) remains constant when monitored over long periods. Amplifiers without 1/f noise have a finite noise level at zero frequency.
- When monitoring low-frequency noise, be careful to keep temperature constant. Temperature drift can look like 1/f noise.
- When using an oscilloscope to measure noise, be sure to use the standard deviation to calculate noise level. The RMS equation includes the DC component. Noise should not contain a DC component.

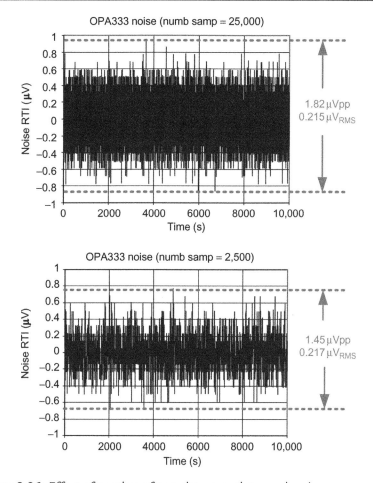

Figure 8.26: Effect of number of samples on peak-to-peak noise measurement

$$RMS = \sqrt{\frac{1}{n}\sum_{i=1}^{n} x_i^2}$$

where
x_i is data samples
n isnumber of samples

$$\sigma = \sqrt{\sigma^2} = \sqrt{\frac{1}{n}\sum_{i=1}^{n} (x_i - \mu)^2}$$

where
x_i is data samples
μ is average of all samples
n is number of samples

Figure 8.27: Standard deviation vs. RMS

Questions

8.1   What frequency does 10 days correspond to?

8.2   What time does $1\,\mu Hz$ correspond to?

8.3   Calculate the total RMS noise for the circuit below after 10 days.

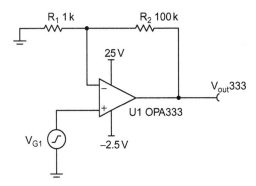

8.4   Calculate the total RMS noise for the circuit below after 10 days

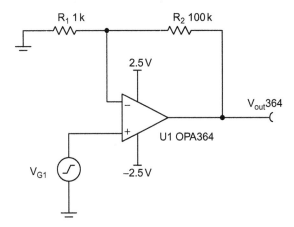

Further Reading

Kugelstadt, T., 2005. Auto-zero amplifiers ease the design of high-precision circuits. Texas Instruments Document Number SLYT204, www.ti.com/lit/an/sly204/sly204.pdf.

Kugelstadt, T., 2007. New zero-drift amplifier has an IQ of $17\,\mu A$. Texas Instruments Document Number SLYT272. http://focus.ti.com/lit/an/slyt272/slyt272.pdf.

CHAPTER 9

Instrumentation Amplifier Noise

This chapter focuses on noise analysis and simulation in instrumentation amplifier circuits. Methods for minimizing noise in instrumentation amplifier design are also discussed.

9.1 Short Review of Three Amp Instrumentation Amplifier

Instrumentation amplifiers (INAs) are used to amplify small differential signals. Most INAs contain several resistors and op-amps. While it is possible to build them using discrete components, there are many advantages of using monolithic integrated circuit INAs. It would be difficult to achieve the accuracy and size of a monolithic INA with discrete components.

Figure 9.1 shows the topology of a three amp INA as well as some of the key connections. The three amp INA is the most popular topology for instrumentation amplifiers. In this section, we develop the gain equation for the INA, which is important for noise analysis. This chapter does not fully explain how to design with and analyze instrumentation amplifiers.

The input signal for an INA is generated by a sensor such as a resistive bridge. To understand the gain equations for an INA, one must first understand the formal definition of the common-mode and differential components in the input signal. The common-mode signal is the average signal on both inputs of the INA. The differential signal is the difference between the two signals. Therefore, by definition, half of the differential signal is above the common-mode voltage and half of the differential signal is below the common-mode voltage. This formal definition of common-mode and differential signals is represented by the signal sources given in Figure 9.2.

Now we will apply the signal source representation of the common-mode and differential voltage developed in Figure 9.2 to a three amp INA and solve for the gain equation. The reason we will go through this exercise is that it will give insight and intuition into our noise analysis. We will simplify the analysis by separating the input stage from the output stage (see Figure 9.3). This will allow us to analyze each half separately so that we may combine them later to achieve the total result.

In Figure 9.4, we begin the analysis by using symmetry to split the upper and lower halves of the input stage. Each half of the amplifier can be seen as a simple noninverting amplifier (with Gain = $R_f/R_{in} + 1$). Note that the gain set resistor is also split in half, so the gain of each half

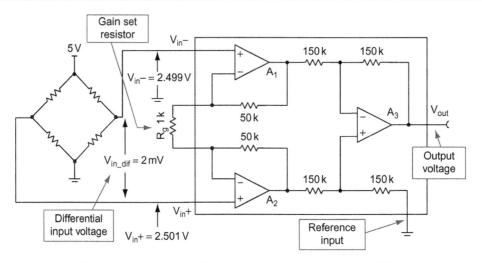

Figure 9.1: Overview of three amp instrumentation amplifier

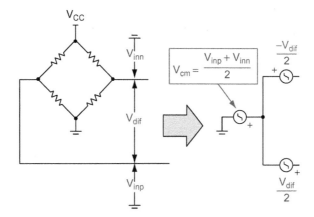

Figure 9.2: Definition of common-mode and differential signal

is Gain = $2R_f/R_g + 1$. Also note that the common-mode voltage (V_{cm}) is transferred to the output of both halves of the amplifier.

Figure 9.5 shows the analysis of the output stage of the INA. This amplifier topology is commonly referred to as a differential amplifier (diff-amp). For the analysis of the output stage, we will break the amplifier in half, analyze both sections, and use superposition to combine the results. The top half of the amplifier is a simple inverting amplifier with a gain of -1 ($V_{out} = -V_{in}$).

The bottom half of the amplifier in Figure 9.5 is a noninverting amplifier with a voltage divider connected to the inputs. Note that the bottom half of the amplifier has two inputs.

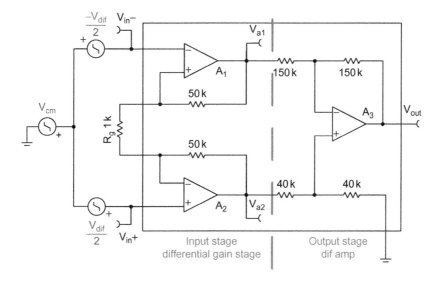

Figure 9.3: Starting the analysis of the three amp INA

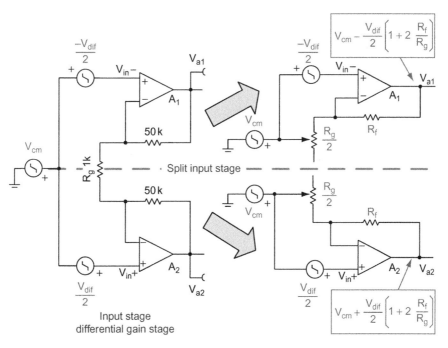

Figure 9.4: Analysis of the three amp INA input stage

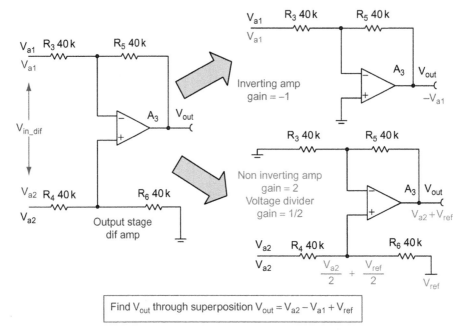

Figure 9.5: Use superposition to analyze the output stage

$$V_{a1} = V_{cm} - \frac{V_{dif}}{2} \cdot \left(1 + 2\frac{R_f}{R_g}\right) \quad \text{(9.1) Input stage top half}$$

$$V_{a2} = V_{cm} + \frac{V_{dif}}{2} \cdot \left(1 + 2\frac{R_f}{R_g}\right) \quad \text{(9.2) Input stage bottom half}$$

$$V_{out} = V_{a2} - V_{a1} + V_{ref} \quad \text{(9.3) Output stage}$$

$$V_{out} = \left[V_{cm} + \frac{V_{dif}}{2} \cdot \left(1 + 2\frac{R_f}{R_g}\right)\right] - \left[V_{cm} - \frac{V_{dif}}{2} \cdot \left(1 + 2\frac{R_f}{R_g}\right)\right] + V_{ref} \quad \begin{array}{l}\text{Substitute}\\ \text{(9.1) and (9.2)}\\ \text{into (9.3)}\end{array}$$

$$V_{out} = V_{dif} \cdot \left(1 + 2\frac{R_f}{R_g}\right) + V_{ref} \quad \text{(9.4) Simplify}$$

Figure 9.6: Solving for the three amp INA transfer function

One input (V_{a1}) is from the input stage, and the other input (V_{ref}) is from the reference pin. The voltage dividers R_4 and R_6 divide both inputs by 2. The gain of the noninverting amplifier is $2(R_5/R_4 + 1)$. The total gain seen by V_{a2} and V_{ref} is 1 (divider gain × noninverting gain = $0.5 \times 2 = 1$).

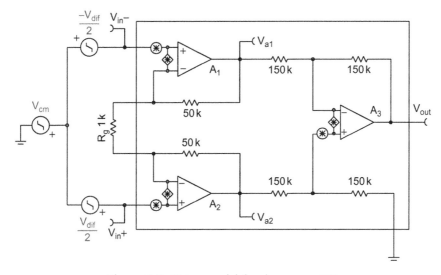

Figure 9.7: Noise model for three amp INA

Combining the results from both halves of the amplifier in Figure 9.5 yields the diff-amp's equation ($V_{out} = V_{a2} - V_{a1} + V_{ref}$). The results from Figures 9.4 and 9.5 are combined for the final transfer function. Note that all the gain is in the first stage; the second stage converts the differential output of the first stage to a single-ended signal. The reference voltage adds directly to the output (gain for the reference signal $= 1$) (Figure 9.6).

9.2 Noise Model of Three Amp Instrumentation Amplifier

Figure 9.7 shows the op-amp noise sources included the INA schematic. Note that each resistor also has a thermal noise associated with it. All of these noise sources can be lumped into as single noise source at the input of the INA or as two noise sources for the input and output stage of the INA. Figure 9.8 shows the simplified noise models with one or two noise sources. Chapter 2 of this book introduces these noise models.

The two-stage model at the top of Figure 9.8 has a voltage noise source for the input stage (V_{n_in}) and a voltage noise source for the output stage (V_{n_out}). The output noise for the entire INA is identified as V_{n_RTO} (voltage noise referred-to-the-output [RTO]). The noise V_{n_RTO} can be computed by adding the root sum of the squares (RSS) of the input noise times the gain and the output noise (see Eq. (9.5) in Figure 9.8). To compute the noise referred-to-the-input (RTI), divide the noise RTO by the gain (see Eq. (9.6) in Figure 9.8).

Figure 9.9 shows the spectral density curve for the INA333. Note that two separate curves are shown for the input stage noise and output stage noise. To use this curve, you need to

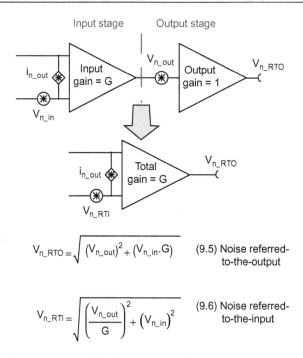

$$V_{n\_RTO} = \sqrt{\left(V_{n\_out}\right)^2 + \left(V_{n\_in} \cdot G\right)} \quad \text{(9.5) Noise referred-to-the-output}$$

$$V_{n\_RTI} = \sqrt{\left(\frac{V_{n\_out}}{G}\right)^2 + \left(V_{n\_in}\right)^2} \quad \text{(9.6) Noise referred-to-the-input}$$

Figure 9.8: Simplified noise model for one or two stages

combine the input stage noise and output stage noise using Eq. (9.5) or (9.6) from Figure 9.8. Eq. (9.5) is included in the spectral density curve for your convenience. The table in Figure 9.9 illustrates how the input stage noise dominates at high gain (e.g., input-referred noise is $50\,\text{nV}/\sqrt{\text{Hz}}$ for gains of 100 and 1000).

Some spectral density curves combine the input stage noise and output stage noise into a single curve. The INA128 noise spectral density curve, shown in Figure 9.10, combines the noise for input and output stage into a single curve. Note that several curves are given at different gains. For low gains, both the input and output stage noise is significant (gains of 1–10). For higher gains, the input noise is dominant (gains of 100–1000).

People sometimes look at the spectral density shown in Figure 9.10 and mistakenly conclude that output noise decreases with gain. Output noise will always increase with larger gain. Therefore, the conclusion is that both the input and output stage contribute noise at low gains, but the input dominates at high gains. Since noise is typically a concern at high gains, the integrated circuit designers optimize the input stage for low noise. It is not as important for the output stage noise to be low, because the input stage normally dominates. The IC designers do not optimize the noise performance of the output stage in order to keep the quiescent current of the amplifier as low as possible (recall from Chapter 6 that noise is inversely proportional to the quiescent current of the amplifier).

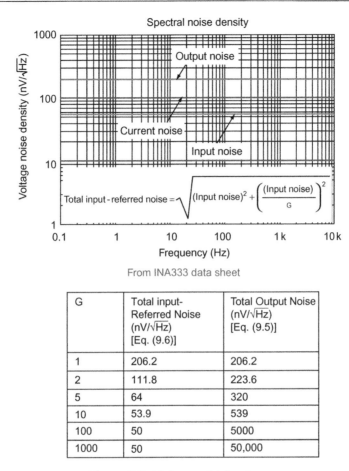

From INA333 data sheet

G	Total input-Referred Noise (nV/√Hz) [Eq. (9.6)]	Total Output Noise (nV/√Hz) [Eq. (9.5)]
1	206.2	206.2
2	111.8	223.6
5	64	320
10	53.9	539
100	50	5000
1000	50	50,000

Figure 9.9: Noise model for three

9.3 Hand Analysis of Three Amp Instrumentation Amplifier

In this section, we compute the expected output noise for a typical INA application. The best way to do this is to analyze the different sections of the circuit separately and combine the results. This analysis will show us which noise sources are significant and which noise sources can be neglected. The ability to determine the dominant noise source is critical in designing a low-noise system; this can save you the effort of attempting to reduce noise in an element that does not have a significant effect on noise performance.

Figure 9.11 shows the example circuit that we will analyze. The gain of this circuit is 101 (Gain = 1 + (100k/1k)). The circuit uses a single supply 5-V instrumentation amplifier. In this example, reference buffer is used to drive the reference pin to half the power supply. This allows the output to swing symmetrically with bipolar input signals. The reference buffer is

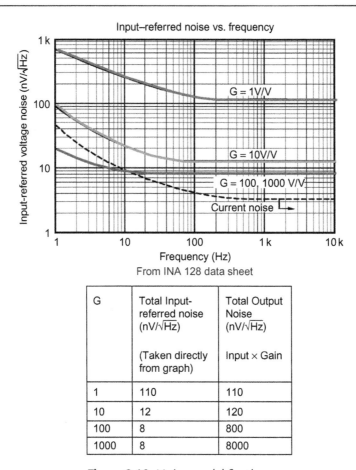

Input–referred noise vs. frequency

From INA 128 data sheet

G	Total Input-referred noise (nV/√Hz)	Total Output Noise (nV/√Hz)
	(Taken directly from graph)	Input × Gain
1	110	110
10	12	120
100	8	800
1000	8	8000

Figure 9.10: Noise model for three

necessary because the reference pin has a relatively high impedance and any series resistance can create a voltage divider error at the noninverting input of A3. The input to this circuit is a bridge sensor. The bridge sensor could measure a wide range of different signals (e.g., pressure, strain, and acceleration). For the purpose of this analysis, however, the bridge sensor is simply modeled as four resistors.

Figure 9.12 shows how the reference buffer noise output is computed. Note that the reference buffer has a voltage divider consisting of two 100-kΩ resistors. From a noise point of view, the two resistors are in parallel (i.e., consider the 5-V supply to be at AC ground potential). To compute the total noise for the reference drive circuit, we consider the thermal noise of the divider, the voltage noise developed from the current of the divider, and the op-amp noise.

Figure 9.13 shows the calculations for the reference buffer from Figure 9.12. The thermal noise from the voltage divider is computed first ($28.7\,\text{nV}/\sqrt{\text{Hz}}$). The voltage noise from the

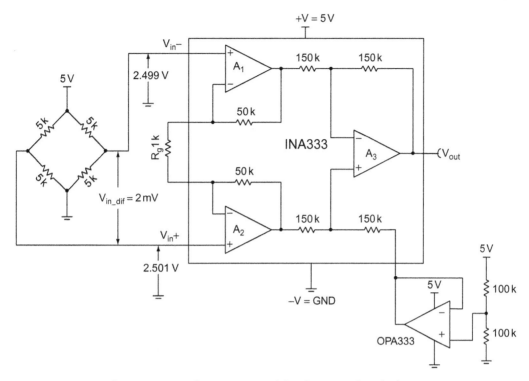

Figure 9.11: Bridge sensor amplifier for example calculation

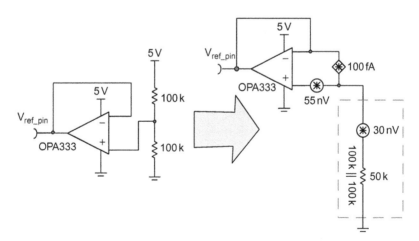

Figure 9.12: Noise equivalent circuit for reference buffer circuit

current noise multiplied by the voltage divider resistance is very small ($5\,\mathrm{nV}/\sqrt{\mathrm{Hz}}$). Typically, current noise from MOSFET op-amps can be neglected unless the input resistance is very large (e.g., greater than $10\,\mathrm{M\Omega}$). The total noise of the reference buffer ($62.2\,\mathrm{nV}/\sqrt{\mathrm{Hz}}$) is dominated by the op-amp noise ($55\,\mathrm{nV}/\sqrt{\mathrm{Hz}}$).

$k_n = 1.38 \times 10^{-23}$ Boltzmann's constant

$T_k = 273 + 25$ Temperature in Kelvin

$R_{eq} = 50\,k\Omega$ $\quad$ Input resistance (parallel combination of voltage divider)

$e_{n\_r} = \sqrt{4k_n \cdot T_n \cdot R_{eq}} = 28.7 \; \dfrac{nV}{\sqrt{Hz}}$ (9.7) Thermal noise from input resistor

$i_n = 100\,fA$ Current noise from OPA333

$e_{n\_i} = i_n \cdot R_{eq} = 5 \; \dfrac{nV}{\sqrt{Hz}}$ (9.8) Voltage noise from current noise

$e_{n-opa} = 55 \; \dfrac{nV}{\sqrt{Hz}}$ $\quad$ Voltage noise from OPA333

$e_{n\_ref} = \sqrt{e_{n\_opa}^2 + e_{n\_r}^2 + e_{n\_i}^2} = 62.2 \; \dfrac{nV}{\sqrt{Hz}}$ (9.9) Total RMS noise from reference driver circuit

Figure 9.13: Reference buffer noise calculations

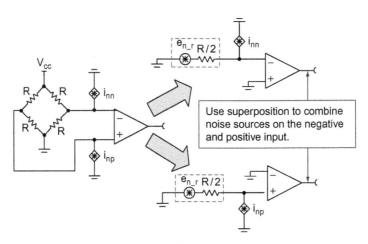

Figure 9.14: Noise model of bridge and input current noise

Therefore, one design consideration for the reference buffer is the thermal noise of the resistor divider. The thermal noise could be reduced significantly by replacing the 100k resistors with 10k resistors. Making this change would reduce the output noise to $55\,nV/\sqrt{Hz}$. The only way to reduce the noise further would be to change the op-amp. Depending on the application, these changes may not be practical. In this example, the

$i_{nn} \cdot \dfrac{R}{2}$ Voltage noise from current noise

$e_{n\_rb} = \sqrt{4k_n \cdot T_n \cdot \dfrac{R}{2}}$ (9.10) Resistor noise

Use superposition to add the noise from the input resistance and both current noise sources.

$$e_{in\_i} = \sqrt{\left(i_{nn} \cdot \dfrac{R}{2}\right)^2 + (e_{n\_rb})^2 + \left(i_{np} \cdot \dfrac{R}{2}\right)^2 + (e_{n\_rb})^2}$$

Assume $\left|i_{nn}\right| = \left|i_{np}\right|$

Note that these sources are uncorrelated

$$e_{in\_i} = \sqrt{2\left(i_n \cdot \dfrac{R}{2}\right)^2 + 2(e_{n\_rb})^2}$$ (9.11) Total noise from input resistors and current source

For this example (R = 5 kΩ, i_n = 100 fA/rt-Hz)

$e_{n\_rb} = 6.4 \dfrac{nV}{\sqrt{Hz}}$ Resistor noise

$i_{nn} \cdot \dfrac{R}{2} = 0.25 \dfrac{nV}{\sqrt{Hz}}$ Voltage noise from current noise

$e_{in\_i} = \sqrt{2(0.5)^2 + 2(9.1)^2} = 9.1 \dfrac{nV}{\sqrt{Hz}}$ Total noise from input resistors and current source

Figure 9.15: Thermal noise and current noise calculation

OPA333 and the voltage divider don't consume much power. Changing the voltage divider and op-amp to improve noise performance would increase the power consumption significantly. Furthermore, we will later see that the noise contribution of the reference buffer is not significant in this circuit.

Figure 9.14 illustrates how to analyze the thermal noise of the resistive bridge sensor and the effect of the current noise. The easiest way to perform this analysis is to consider each input separately and to use superposition to combine the noise from each input. V_{cc} acts as an AC ground, so each input sees two legs of the bridge resistance in parallel. Thus, each input sees one half of the bridge resistance (R/2) as an equivalent input resistance.

Figure 9.15 shows the calculations for the thermal noise from the bridge sensor and the effect of the current noise. The current noise is extremely low in this example because the INA333 is a MOSFET instrumentation amplifier ($100\,fA/\sqrt{Hz}$). The current noise is multiplied by the equivalent input resistance to convert it to a voltage noise.

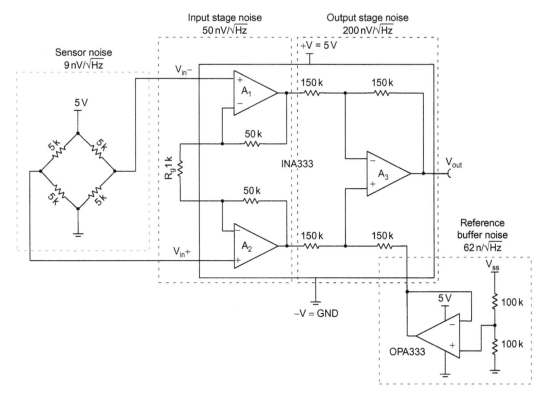

Figure 9.16: Summary of noise components in example circuit

Figure 9.16 shows all the noise components in our example circuit. Note that the input stage noise and the sensor noise add as the RSS. The combined input and sensor noise is multiplied by the INA gain and RSS added with the output stage noise and reference buffer noise.

In many cases when doing noise analysis, you have a single dominant source of noise and all other noise sources can be neglected without introducing significant error. A good rule of thumb is that a noise source is dominant if it is three times larger than the other sources. Keep in mind that noise is added as the RSS, so the factor of three is squared. Figure 9.17 illustrates the "rule of three."

For our example circuit, you can see that the input stage noise dominates. Remember to use the "rule of three" when comparing the different components with each other. The input stage noise is $50\,\text{nV}/\sqrt{\text{Hz}}$ and the total combined noise RTI is $50.8\,\text{nV}/\sqrt{\text{Hz}}$. It is important to identify the dominant noise components so that you don't waste time trying to optimize the noise of circuits that do not matter. In this example, the only important factor to overall noise is the input stage noise. Thus, optimizing the reference buffer noise would not yield a measurable benefit (Figure 9.18).

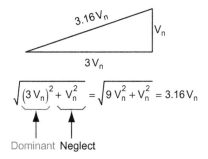

$$\underbrace{\sqrt{\left(3\,V_n\right)^2 + V_n^2}}_{} = \underbrace{\sqrt{9\,V_n^2 + V_n^2}}_{} = 3.16\,V_n$$

Dominant Neglect

Figure 9.17: Rule of three identifies dominant noise components

(9.12) Noise spectral density referred-to-the-input for this example

$$\text{Noise\_Spec\_Den\_RTI} = \sqrt{V_{n\_in\_stage}^2 + V_{n\_bridge}^2 + \left(\frac{V_{n\_out\_stage}}{G}\right)^2 + \left(\frac{V_{n\_ref\_buf}}{G}\right)^2}$$

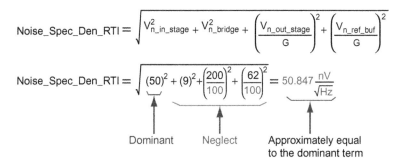

$$\text{Noise\_Spec\_Den\_RTI} = \sqrt{(50)^2 + (9)^2 + \left(\frac{200}{100}\right)^2 + \left(\frac{62}{100}\right)^2} = 50.847\,\frac{nV}{\sqrt{Hz}}$$

Dominant Neglect Approximately equal
to the dominant term

Figure 9.18: Input noise dominates in this example

The next step in the analysis is to compute the total RMS noise and peak-to-peak noise at the output. To compute RMS noise, we must know the noise bandwidth. The gain vs. frequency plot from the data sheet can be used to estimate the order of the roll-off of high-frequency gain. In Figure 9.19, you can see that gain rolls off at $-20\,\text{dB/decade}$ (equivalent to a first-order or single-pole filter). The roll-off is used to compute the brick wall correction factor (K_n). The brick wall correction factor is multiplied by the 3-dB bandwidth to compute the noise bandwidth. In this example, $K_n = 1.57$ because the INA333 high-frequency roll-off is single pole. The 3-dB bandwidth can be read directly from the "frequency response" table (3.5 kHz). Thus, for this example, the noise bandwidth is $BW_n = 1.57 \times (3.5\,\text{kHz}) = 5.495\,\text{kHz}$. Chapter 2 of this book discusses the brick wall correction factor and noise bandwidth.

Figure 9.20 shows the final calculations for the example circuit. Note that the method used to compute total noise is described in Chapter 2 of this book. This example uses a chopper-stabilized instrumentation amplifier. Thus, the amplifier does not have any 1/f noise. This simplifies the

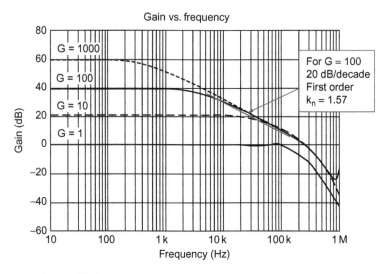

Parameter	INA333			Unit
	Min	Typ	Max	
Frequency response				
G = 1		150		kHz
G = 10		35		kHz
G = 100		3.5		kHz
G = 1000		350		Hz

Figure 9.19: Noise model for three

calculations. The RMS noise is calculated by taking the square root of the noise bandwidth multiplied by the input noise (see Eq. (9.13) in Figure 9.20). Finally, the peak-to-peak voltage can be approximated using six times the RMS value (see Eq. (9.14) in Figure 9.21).

9.4 Simulation of Three Amp Instrumentation Amplifier

The circuit from Figure 9.11 can be simulated using any Spice simulator. "TINA Spice" can be downloaded from the Texas Instruments web page at no cost. Furthermore, the model for INA333 can be downloaded. The INA333 model correctly models noise, and most other parameters of interest. Chapter 4 of this book shows how to simulate noise in op-amp circuits; the same methods are used for INAs.

Figure 9.21 shows the "noise analysis" options available from TINA Spice. When the "Output Noise" box is checked, the simulator will create a spectral density plot at each test point in the schematic. When the "Total Noise" box is checked, the simulator will generate a plot of the RMS noise (i.e., the power spectral density plot integrated).

$G = 100$ INA gain

$V_{in\_RTI} = 50.85 \, nV/\sqrt{Hz}$ From "input referred noise" equation

$f_H = 3.5 \, kHz$ From data sheet table for gain $= 100$

$K_n = 1.57$ For first-order function
See gain vs. frequency in the data sheet

$BW_n = f_H \cdot K_n = 5.495 \, kHz$ Noise bandwidth

$e_{n\_out} = G \cdot V_{in\_RTI} \cdot \sqrt{BW_n} = 376.9 \, \mu VRMS$ (9.13) RMS output noise

$e_{n\_outpp} = 6 \cdot e_{n\_out} = 2.26 \, mVpp$ (9.14) Peak-to-peak output

Figure 9.20: Final noise calculation for bridge resistor circuit

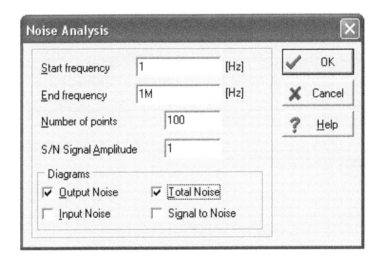

Figure 9.21: TINA noise analysis options

Figure 9.22 shows the simulated spectral density plot for the example circuit. This is generated in TINA Spice using the "Output Noise" option. Figure 9.23 shows the RMS noise (square root of power spectral density integrated).

Note that the simulated and calculated RMS noise in Figure 9.23 does not match exactly (simulated $= 422 \, \mu V_{RMS}$, calculated $= 377 \, \mu V_{RMS}$). There are several reasons for this minor discrepancy. First, the spectral density of the simulated result is slightly higher than the calculated spectral density. Second, the bandwidth for the simulated result is slightly wider than the bandwidth from the data sheet. Finally, the simulator gain roll-off has a "bump" at

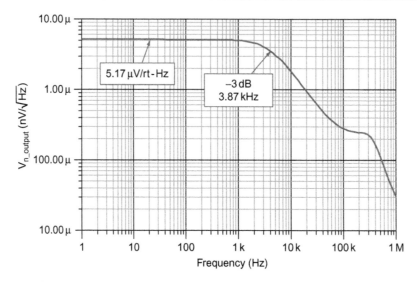

Figure 9.22: Spectral density at output of INA333 example circuit

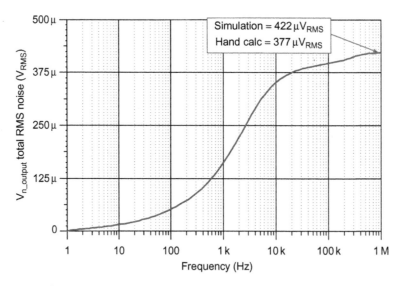

Figure 9.23: RMS output noise for INA333 example circuit

approximately 200 kHz. This "bump" was not accounted for in the hand calculations. These discrepancies are illustrated in Figure 9.24.

Engineers are often concerned when they see discrepancies between simulations and hand calculations. In this case, the error is of the order of 10%. A 10% error on noise calculations is actually not very large. Remember that the specifications are given as typical, so you will

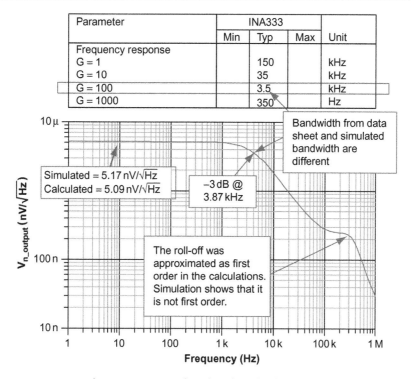

Parameter	INA333			
	Min	Typ	Max	Unit
Frequency response				
G = 1		150		kHz
G = 10		35		kHz
G = 100		3.5		kHz
G = 1000		350		Hz

Figure 9.24: Simulated vs. hand calculations

see variations greater than 10% for real-world devices. In this example, the hand calculations are more accurate than the simulated results for bandwidth and spectral density below 1 kHz, because they are based directly on the data sheet. The simulation more accurately follows the data sheet at the 200-kHz bump in the spectral density curve. In general, discrepancies of this order should be ignored.

9.5 Reducing Noise with Averaging Circuit

One way to reduce noise is to connect several amplifiers' inputs together and average the outputs using an op-amp averaging circuit. Figure 9.25 shows an inverting averaging circuit. All of the input resistors (R_1, R_2, R_3, ..., R_N) must be equal for the averaging feature to operate. Also, the feedback resistor (R_f) must be equal to the input resistor value divided by the number of input resistors.

Figure 9.26 mathematically shows how the averaging circuit reduces noise. The averaging effect assumes that the noise sources are equal and uncorrelated. The output noise is the input noise divided by the square root of the number of averages.

Figure 9.27 shows a practical circuit averaging three INA333 amplifiers. Note that the INA333 gain is set to 1001 in this example (Gain = 1 + (100k/100)). Note also that the inputs

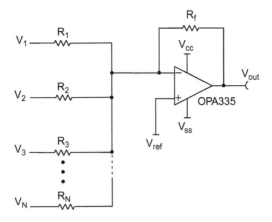

$$V_{out} = V_{ref} - R_f \left(\frac{V_1}{R_1} + \frac{V_2}{R_2} + \frac{V_3}{R_3} + ... + \frac{V_N}{R_N} \right) \qquad (9.15)$$

For an averaging circuit, choose

$R_1 = R_2 = R_3 = ... = R_N = R$

$R_f = R/N$

$$V_{out} = V_{ref} - \frac{(V_1 + V_2 + V_3 + ... + V_N)}{N} \qquad (9.16)$$

Figure 9.25: Op-amp averaging circuit

$$V_{noise\_output} = \sqrt{\left(\frac{V_{noise1}}{N} \right)^2 + \left(\frac{V_{noise2}}{N} \right)^2 + \left(\frac{V_{noise3}}{N} \right)^2 + ... + \left(\frac{V_{noiseN}}{N} \right)^2}$$

where V_{noise1}, V_{noise2}, V_{noise3}, ..., V_{noiseN} are noise sources.

If you assume that V_{noise1}, V_{noise2}, V_{noise3}, ..., V_{noiseN} are equal uncorrelated noise sources, then

$$V_{noise\_output} = \sqrt{N \left(\frac{V_{noise}}{N} \right)^2} = \sqrt{\frac{V_{noise}^2}{N}} = \frac{V_{noise}}{\sqrt{N}} \qquad (9.17)$$

Figure 9.26: Derivation of output noise for averaging circuit

of the three amplifiers are connected together. The outputs of the three INA333 are connected to the averaging circuit. This circuit effectively acts as a single INA333 with noise reduced by the factor of 1/sqrt(3). Note that the current noise from the INA333 input amplifiers adds (root sum of squares). Thus, this circuit is an effective way to decrease voltage noise, but actually increases current noise.

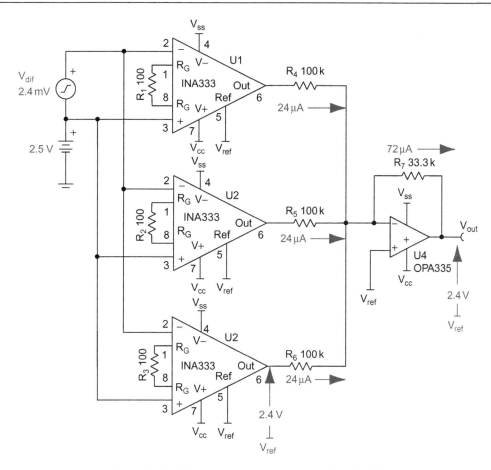

Figure 9.27: Three amp averaging circuit for INA333

The value of R_4, R_5, and R_6 was selected to limit current to a reasonable value. The INA333 and OPA335 are low-power devices and care must be taken to prevent excessive current flow. Engineers sometimes choose a smaller value for averaging circuit input resistors to minimize noise; however, this resistor value will not affect the overall noise because the INA333 noise dominates ($50\,\text{nV}/\sqrt{\text{Hz}} \times 1000 = 50\,\mu\text{V}/\sqrt{\text{Hz}}$). R_7 is selected to scale the gain for averaging (e.g., $100\,\text{k}/3 = 33\,\text{k}$).

Figure 9.28 shows the test board used to prove the effectiveness of the averaging circuit. The number of amplifiers that are averaged is jumper selectable. The instrumentation amplifier gain is set using socketed through-hole resistors. The goal was to make the board flexible enough to do a wide range of experiments. Using the jumpers and through-hole components makes the size of the printed circuit board much larger than it would be with only surface mount components. The size is important because the longer traces are more likely to pick up extrinsic noise (e.g., 60 Hz and RFI interference).

Figure 9.28: Circuit used to demonstrate effect of averaging on noise

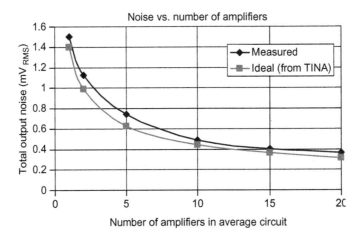

Figure 9.29: Noise vs. number of amplifiers for INA333 circuit

Figure 9.29 shows the total output noise vs. the number of amplifiers averaged. The noise decreases according to the equation $V_{n\_output} = V_{noise}/sqrt(N)$. Note that a significant noise reduction is made using a four amplifier circuit (e.g., four amplifiers reduce noise by one-half (1/sqrt (4) = ½)). Using a large number of amplifiers further reduces noise, but this may not be practical because of PC board area, cost, and complexity.

Figure 9.30 shows the measured and simulated spectral density for INA333 averaging circuit. The measured and simulated results compare closely with each other.

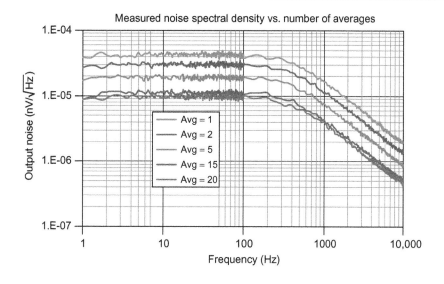

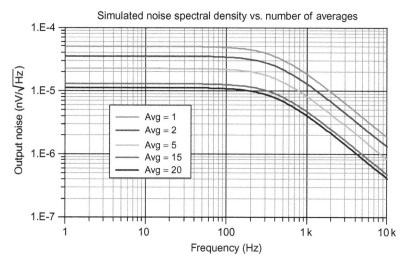

Figure 9.30: Spectral density vs. number of averages

Chapter Summary

- Instrumentation amplifier noise can be specified in two ways:
  - All the noise from both the input and output stage can be referred to the input.
  - Both the input and output noise can be specified.
- A common misconception with instrumentation amplifiers is that noise will decrease with gain.
  - Output noise will always increase with gain.

- Input-referred noise is the output noise divided by gain. For low gain, both the input and output stage contribute to total noise. For high gain, the input noise will dominate. Thus, the input-referred noise is smaller for high gains than low gains.
- When doing a noise analysis on a complex system, try to identify the dominant noise source.
  - Knowing the dominant noise source helps you to know which circuit element to focus noise reduction efforts on.
  - In the noise example given in this chapter, the input stage noise of the amplifier was the dominant noise source.
- Three common ways to reduce noise are to limit bandwidth, minimize resistor values, and choose low-noise devices.
  - Noise can also be reduced with averaging.

Questions

9.1   Calculate the total output noise for the circuit below.

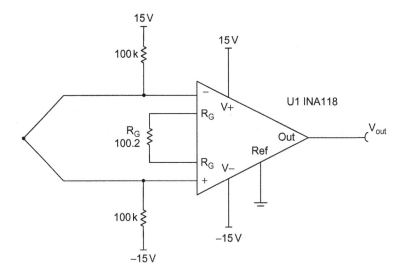

9.2   The INA333 is a zero-drift amplifier and so it has no 1/f noise. The INA331 is not a zero-drift amplifier and so it has 1/f noise. For low bandwidth the INA333 is a better solution for low noise, and for higher bandwidth the INA331 is a better low-noise solution. Find the bandwidth (in Hz) where the noise contribution of the two amplifiers is equal.

9.3   What is the total output noise for the circuit below?

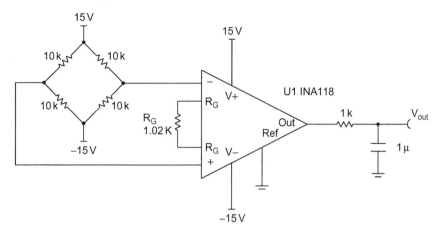

Further Reading

Hann, G., 2008. Selecting the right op amp—Electronic Products. *Electronic Products Magazine–Component and Technology News* 21 November 2008. <http://www2.electronicproducts.com/Selecting_the_right_op_amp-article-facntexas_nov2008-html.aspx/> (accessed 9.10.2009).

CHAPTER 10

Photodiode Amplifier Noise

This chapter focuses on noise analysis and simulation in photodiode amplifier circuits. In addition, the process for optimizing the feedback compensation capacitor for minimization of output noise is a key focus.

10.1 Introduction to Photodiodes

To understand how to properly configure and analyze a photodiode in an amplifier configuration, it is important to understand some basic fundamentals on photodiode operation. A photodiode is a semiconductor device that is used to convert light to electrical current or voltage. Figure 10.1 shows a simple PN photodiode consisting of N- and P-doped semiconductor material. With no bias applied to the diode, the free electrons from the N region combine with the free holes in the P region to create a depletion region. The depletion region is charged positively in the N material and negatively in the P material, so it develops an e-field.

The schematic shown in Figure 10.2 is the model for a photodiode. The different components in the model are normally given in the photodiode data sheet. The junction capacitance C_j, shunt resistance R_{sh}, and dark current I_D are key parameters used in noise analysis.

The purpose of a photodiode is to convert light to a proportionate current. Figure 10.3 shows that the responsivity of a photodiode to light is affected by the wavelength. Different types of photodiodes are designed and optimized to respond at specific wavelengths.

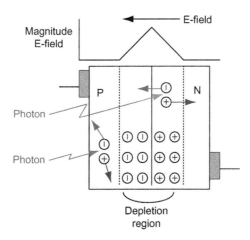

Figure 10.1: Simplified semiconductor model of photodiode

Operational Amplifier Noise.

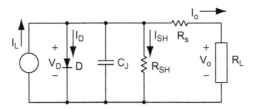

I_L is current generated by the incident light
I_D is diode dark current
C_j is junction capacitance
R_{sh} is shunt resistance
R_s is series resistance
I_{SH} is shunt resistance current
V_D is diode voltage
I_o is output current
V_o is output voltage

Figure 10.2: Photodiode electrical model

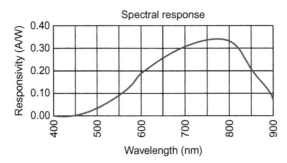

I_L is current generated by the incident light
$I_L = r_\phi \phi_e$
r_ϕ is the diode's flux responsivity
ϕ_e is the radiant flux energy in Watts

Figure 10.3: Photodiode responsivity vs. wavelength

The junction capacitance is a key parameter used in noise analysis. Typically, smaller junction capacitance will lead to lower noise. Increasing the reverse bias voltage on the diode will decrease the junction capacitance. In some applications, the reverse bias voltage is increased to reduce the junction capacitance (Figure 10.4).

Figure 10.5 shows how the diode characteristic curves are shifted by applied light. With no light applied, the photodiode acts as a conventional rectifier. Increasing applied light will shift the curve downward on the current axis.

C_j is junction capacitance

$$C_j = \frac{C_{j0}}{\sqrt{1 + \dfrac{V_R}{\phi_B}}}$$

C_{j0} is the photodiode capacitance at zero bias

ϕ_B is the built-in voltage of the diode junction

V_R is the reverse bias voltage

Figure 10.4: Photodiode junction capacitance vs. reverse bias voltage

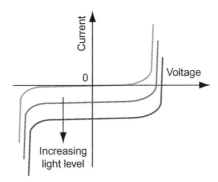

Figure 10.5: Photodiode V–I characteristics

10.2 The Simple Transimpedance Amplifier

The circuit shown in Figure 10.7 is the most common photodiode amplifier. Because it converts current to voltage, it is called a transimpedance amplifier. This chapter focuses on the simple transimpedance topology. The output voltage is calculated by multiplying the input current from the photodiode by the feedback resistor R_f.

10.2.1 Bandwidth for Simple Transimpedance Amplifier

In this chapter, we focus on the noise bandwidth and noise gain of the transimpedance amplifier. However, it is also important to understand the bandwidth limit for the photodiode signal (signal bandwidth). The signal bandwidth (f_p) is shown in Figure 10.6. Note that the signal bandwidth is limited by R_f and C_f.

10.2.2 Noise Model for Simple Transimpedance Amplifier

Analysis of the photodiode amplifier noise can be broken into three subsections: the photodiode, the op-amp, and the resistor. Noise sources from all three subsections will be combined to compute the total noise. The input capacitance (C_{in}), feedback capacitance (C_f) and feedback resistance (R_f) in this circuit will significantly affect the noise because they shape the noise-gain curve.

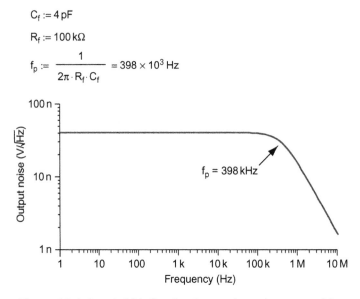

$C_f := 4\,pF$

$R_f := 100\,k\Omega$

$f_p := \dfrac{1}{2\pi \cdot R_f \cdot C_f} = 398 \times 10^3\,Hz$

$f_p = 398\,kHz$

Figure 10.6: Bandwidth for simple transimpedance amplifier

10.3 Photodiode Current Noise

The current noise of a photodiode is the root sum square of three different noise sources: the thermal (Johnson) noise of the shunt resistance, the dark current shot noise, and the shot noise of the light current. Normally, we consider the thermal noise of a resistor to be a voltage noise. However, in the case of photodiodes, it is convenient to consider the thermal noise as a current.

Photodiode noise also contains shot noise. Shot noise is proportional to DC current flow and is only present when DC current flows. Two types of shot noise are present in the photodiode circuit. One is caused by the current that flows when light is applied to the photodiode (I_L). The other noise source is caused by the dark current (I_D). Figure 10.5 shows the photodiode shot noise equations.

The three current sources from the diode can be added using the root sum of the square (see Figure 10.5). The total noise current will flow through the feedback resistor R_f. Figure 10.7 shows how the current noise converts to an RMS voltage noise at the output. The bandwidth limit for the current noise is equivalent to the signal bandwidth of the transimpedance amplifier multiplied by the brick wall correction factor (noise bandwidth).

10.4 Thermal Noise from R_f

The thermal noise of the feedback resistor can be calculated using the equations shown in Figure 10.5. The bandwidth limit for the thermal noise is equivalent to the signal bandwidth of the transimpedance amplifier multiplied by the brick wall correction factor (noise bandwidth) (Figures 10.6 and 10.8–10.13).

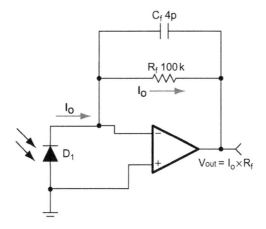

Figure 10.7: Common transimpedance amplifier

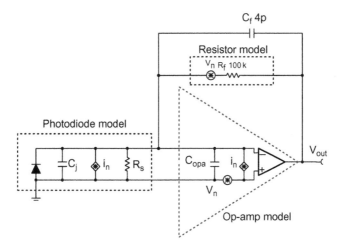

Figure 10.8: Noise model for simple transimpedance amplifier

$$i_j = \sqrt{\frac{4_{kb} \cdot T_n}{R_{sh}}} \text{ Thermal (Johnson noise)}$$

k_b is Boltzmann constant 1.38×10^{-23}J/K

T_n is temperature in Kelvin (25 °C)

R_{sh} is shunt resistance in photodiode

Figure 10.9: Photodiode thermal current noise spectral density

$$i_{sD} = \sqrt{2q \cdot I_D}$$ Shot noise (dark)

$$i_{sL} = \sqrt{2q \cdot I_L}$$ Shot noise (with light)

q is electron charge 1.6×10^{-19} °C

I_D is dark current in photodiode

I_L is photo current in photodiode

Figure 10.10: Photodiode shot current noise spectral density

$$i_{n\_diode} = \sqrt{i_j^2 + i_{sD}^2 + i_{sL}^2}$$ Total diode current noise

i_j is thermal (Johnson noise)

i_{sD} is shot noise current from dark current

i_{sL} is shot noise current from photo current

Figure 10.11: Total photodiode current noise spectral density

$$i_{n\_total} = \sqrt{i_{n\_opa}^2 + i_{n\_diode}^2}$$ Total noise current

$$BW_n = K_n \cdot f_p$$ Noise bandwidth (brick wall filter)

$$E_{nol} = i_{n\_total} \cdot R_f \cdot \sqrt{BW_n}$$ Voltage noise at output from total current noise

Figure 10.12: RMS noise voltage from photodiode current noise

$$E_{noR} = \sqrt{4k_b \cdot T_n \cdot R_f \cdot BW_n}$$ Thermal noise at output

$$BW_n = K_n \cdot f_p$$ Noise bandwidth (brick wall filter)

$$k_b = 1.38 \times 10^{-23} \frac{J}{K}$$ Boltzmann constant

T_n is temperature in Kelvin

f_p is transconductance bandwidth

K_n is brick wall factor

Figure 10.13: Thermal noise from feedback resistor in transimpedance amplifier

10.5 Noise from Op-amp Voltage Noise Source

The final contribution to the output noise to consider is the op-amp noise voltage source (V_n in Figure 10.14). The calculation of this contribution is complicated by the AC noise gain of the amplifier. Recall that the noise gain is the gain seen by the noise voltage signal source. In this example, C_f and C_{in} will cause a peak in the noise-gain curve that will significantly affect the total output noise. The circuit shown in Figure 10.14 shows the key components that affect the noise gain.

To compute the noise gain for the photodiode amplifiers, we will perform a nodal analysis at the summing junction of the inverting amplifier input. This analysis yields for three current paths (see Figure 10.15). To complete this analysis, we need to consider the impedance of the capacitors. In this case, we will represent the impedance of a capacitor as $1/(sC_f)$, where s is jω.

Figure 10.16 shows the algebra for the nodal analysis. The first equation in this figure shows the equation taken directly from the node. Note that the equation has three terms for each of the three current paths. The first equation is rearranged into the noise-gain transfer function. The numerator

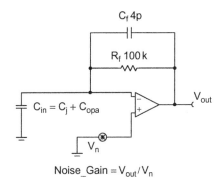

Figure 10.14: Noise from op-amp voltage noise source

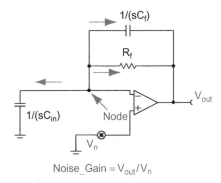

Figure 10.15: Nodal analysis for noise gain vs. frequency

Nodal analysis on transimpedance amp

$$\frac{V_n}{\dfrac{1}{s \cdot C_{in}}} + \frac{(V_n - V_{out})}{R_f} + \frac{V_n - V_{out}}{\dfrac{1}{s \cdot C_f}} = 0$$

Solve for noise gain V_{out}/V_n

$$\frac{V_{out}}{V_n} = \frac{R_f \cdot (C_f + C_{in}) \cdot s + 1}{C_f \cdot R_f \cdot s + 1}$$

The numerator contains a zero

$$f_z = \frac{1}{2\pi \, R_f \cdot (C_f + C_{in})}$$

The denominator contains a pole

$$f_p = \frac{1}{2\pi \, R_f \cdot C_f}$$

Figure 10.16: Solutions for pole and zero using nodal analysis

$$f_i = \frac{C_f}{C_i + C_f} \cdot f_c \quad \begin{array}{l}\text{Intersection of the noise gain curve}\\ \text{with the AOL curve}\end{array}$$

f_c is Unity gain bandwidth from op-amp data sheet

$$GPM = 1 + \frac{C_{in}}{C_f} \quad \text{Gain peak magnitude}$$

Figure 10.17: Key equations for the noise-gain curve

of this transfer function contains a zero and the denominator contains a pole. The zero (f_z) and pole (f_p) will have significant effects on the transfer function. The equations for the pole and zero are shown at the bottom of Figure 10.16.

Noise gain vs. frequency is displayed in Figure 10.18. Note that noise gain is 0 dB until the zero (f_z). Between the zero (f_z) and the pole (f_p), the noise gain will rise at 20 dB/decade. The pole and zero cancel so that the noise gain flattens out. The flat region continues until it intercepts the AOL curve at f_i and then rolls off with AOL. The magnitude of the flat region between f_p and f_i is dependent on C_{in} and C_f. The flat region between f_p and f_i is called the noise-gain peak. Each frequency transition point in Figure 10.7 will be useful in deriving equations for the total noise.

Figure 10.17 gives some of the key equations from the noise-gain curve shown in Figure 10.18. The equation for f_i shows where the noise-gain curve intercepts the AOL curve. The equation for gain peak magnitude (GPM) gives the magnitude of the noise-gain peak in deabels.

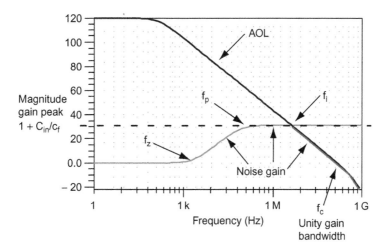

Figure 10.18: Noise gain vs. frequency

Recall that noise gain is the gain seen by the op-amp to its noise voltage source. Figure 10.19 shows the op-amp voltage noise curve at the top of the figure, the noise gain in the centre of the figure, and the resultant output noise at the bottom of the figure. The key here is to understand that the op-amp noise curve is multiplied by the noise-gain curve to produce the output noise curve. To find the total RMS output noise we must integrate the output noise curve.

Figure 10.20 shows a circuit that can be used to generate the noise gain, AOL, and current-to-voltage gain. A Spice AC sweep is used to generate curves at the test points V_{F1}, V_{F2}, and V_{F3}. These signals are postprocessed using the formulas from Figure 10.21. Note that the 1-TH inductor is used to break the feedback loop from an AC perspective but allow for a DC connection. The 1-TF capacitor allows the signal source V_{G1} for AC coupling into the loop at extremely low frequencies. These values are not practical for any physical circuit, but work well for this Spice curve generation technique.

Figure 10.21 shows how to start an "AC Transfer Characteristic" using TINA Spice. After starting the AC Transfer Characteristic, you will need to enter the Start Frequency and End Frequency as required for your application. The AC Transfer Characteristic will create a curve for each of the test points and meters in the circuit.

After creating the AC Transfer Characteristic, you will have to use the postprocessing capability to create the key transimpedance curves. Postprocessing allows you to do mathematics on the curves generated by the AC Transfer Characteristic. Figure 10.21 shows the formulas that are used to generate the key transimpedance plots. The equations are entered into the postprocessor (Figure 10.22).

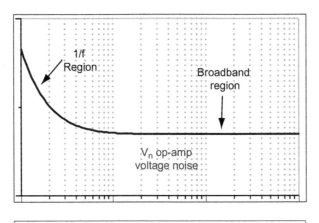

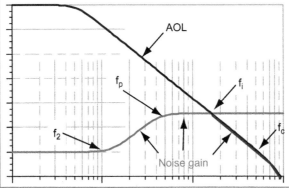

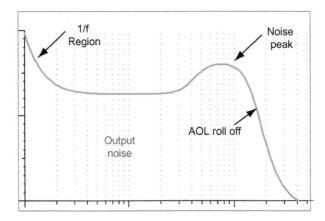

Figure 10.19: Output noise is input noise × noise gain

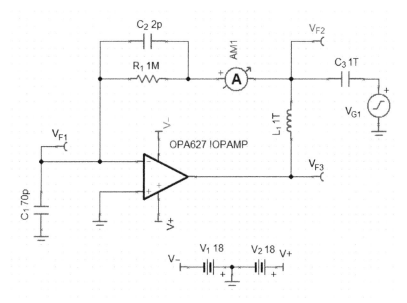

Figure 10.20: Spice circuit to find AOL and noise gain

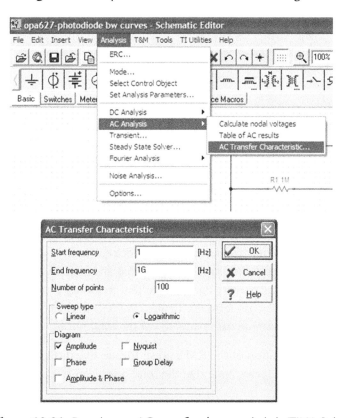

Figure 10.21: Running an AC transfer characteristic in TINA Spice

$$AOL = \frac{V_{F3}}{V_{F1}}$$

$$Noise\_Gain = \frac{1}{\beta} = \frac{V_{F2}}{V_{F1}}$$

$$I\_to\_V\_Gain = \frac{V_{F3}}{AM1}$$

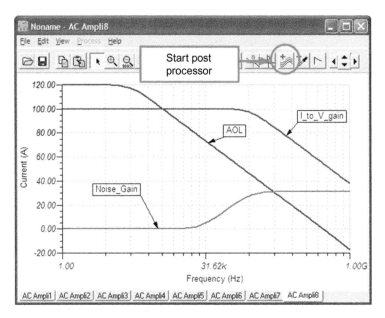

Figure 10.22: Noise gain and AOL in AC plot

Figure 10.23 shows the postprocessor. The AOL, noise-gain, and current-to-voltage curves are generated by entering the equation into the postprocessor. The example shows the current-to-voltage equation being entered.

The output noise spectral density can be computed by multiplying the noise-gain curve by the voltage noise spectral density. The output noise spectral density curve and equation are shown in Figure 10.24. This equation has four terms that are each dominant during a specific frequency region in the spectral density curve. Note that Region 2 is a constant region between Regions 1 and 3, where no term dominates. We will integrate the power spectral density in each region and combine the results to get the total noise.

Figure 10.25 shows the derivation of the RMS noise for the five different regions of the output noise spectral density curve. Note that the total noise can be computed using the root sum of the square of each noise component.

Figure 10.26 shows the output noise spectral density with a logarithmic and linear *x*-axis. The linear scale is intended to emphasize that Regions 3, 4, and 5 dominate the total noise. When

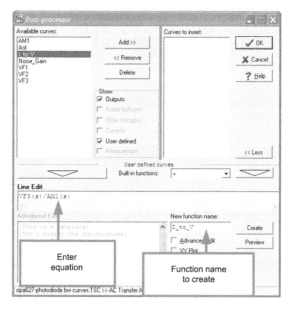

Figure 10.23: Creating noise gain and AOL with postprocessor

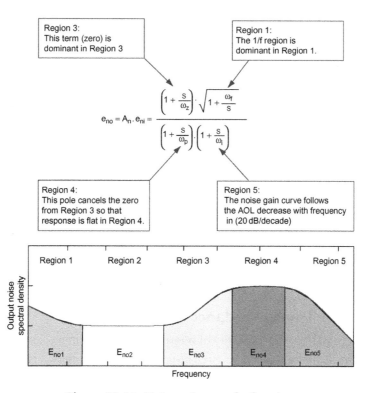

Region 3:
This term (zero) is dominant in Region 3

Region 1:
The 1/f region is dominant in Region 1.

$$e_{no} = A_n \cdot e_{ni} = \frac{\left(1 + \dfrac{s}{\omega_z}\right) \cdot \sqrt{1 + \dfrac{\omega_f}{s}}}{\left(1 + \dfrac{s}{\omega_p}\right) \cdot \left(1 + \dfrac{s}{\omega_l}\right)}$$

Region 4:
This pole cancels the zero from Region 3 so that response is flat in Region 4.

Region 5:
The noise gain curve follows the AOL decrease with frequency in (20 dB/decade)

Region 1   Region 2   Region 3   Region 4   Region 5

Output noise spectral density

E_{no1} E_{no2} E_{no3} E_{no4} E_{no5}

Frequency

Figure 10.24: Noise-gain transfer function

$$E_{noe1} = \sqrt{\int_{f_L}^{f_f} \frac{e_{nif}^2 f_f}{f} \, df} = e_{nif} \cdot \sqrt{f_f \cdot 1n\left(\frac{f_f}{f_L}\right)}$$

$$E_{noe2} = \sqrt{\int_{f_f}^{f_z} e_{nif}^2 \, df} = e_{nif} \cdot \sqrt{(f_z - f_f)}$$

$$E_{noe3} = \sqrt{\int_{f_z}^{f_p} \frac{e_{nif}^2 f^2}{f_z^2} \, df} = \frac{e_{nif}}{f_z} \cdot \sqrt{\frac{f_p^3 - f_z^3}{3}}$$

$$E_{noe4} = \sqrt{\int_{f_p}^{f_i} e_{nif}^2 \cdot \left(\frac{C_{in} + C_f}{C_f}\right)^2 df} = e_{nif} \left(\frac{C_{in} + C_f}{C_f}\right) \cdot \sqrt{(f_i - f_p)}$$

$$E_{noe5} = \sqrt{\int_{f_i}^{\infty} \frac{e_{nif}^2 \cdot f_c^2}{f^2} \, df} = \frac{e_{nif} \cdot f_c}{\sqrt{f_i}}$$

$$E_{noe} = \sqrt{E_{noe1}^2 + E_{noe2}^2 + E_{noe3}^2 + E_{noe4}^2 + E_{noe5}^2}$$

Figure 10.25: Total RMS noise from noise voltage in transimpedance amplifier

looking at the log scale, it is easy to be misled into believing that the 1/f region (R1) could be the dominant source of noise. The linear scale helps give a better view of the dominant region.

10.6 Total Noise (Op-amp, Diode, and Resistance)

At this point, we have determined relationships for all three noise sources in the transimpedance circuit: resistor noise, current noise, and voltage noise. To compute the total output noise, we combine these three results using the root sum of the square (Figure 10.27).

10.7 Stability of Transimpedance Amplifier

In addition to understanding noise, stability of transimpedance amplifiers is an important concern. Also, understanding the noise-gain curve gives insight into the amplifiers' stability. Large capacitance on the input of an op-amp tends to make it unstable. Because photodiodes have large

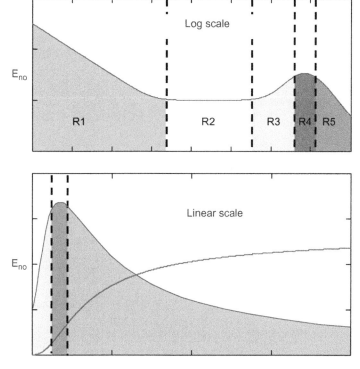

Figure 10.26: Different regions in voltage noise curve on logarithmic and linear scales

$$E_{no} = \sqrt{E_{noR}^2 + E_{noI}^2 + E_{noe}^2}$$ Total output noise for the transimpedance amp

Figure 10.27: Total RMS output noise for transimpedance amplifier

capacitance, they inherently have stability issues. Figure 10.28 shows a photodiode amplifier without a feedback capacitance. We will prove through analysis that this topology is unstable.

Figure 10.29 shows the noise-gain and AOL curve for Figure 10.28. The noise gain continuously increases by 20 dB/decade after the zero because there is no pole from a feedback capacitor. Rate of closure is defined as the difference between the slope of the AOL curve and the noise-gain curve at the point where they intersect. The rate of closure is 40 dB/decade in Figure 10.20 because the AOL curve is decreasing by 20 dB/decade and the noise gain is increasing by 20 dB/decade. To ensure stability, the rate of closure should be 20 dB/decade or less. So the example in Figure 10.20 is not stable because its rate of closure is 40 dB/decade.

Another approach to testing stability is to apply a step input to a circuit. For this example, a step input of current (light) was applied. You can see that the output has substantial ringing

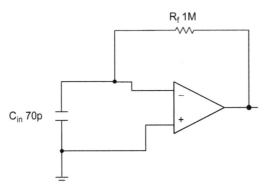

Figure 10.28: Transimpedance amplifier without C_f is not stable

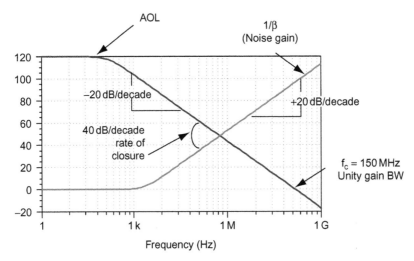

Figure 10.29: Stability test fails because the rate of closure is 40dB/decade

and is not stable. In practical conditions, it is possible that this circuit would oscillate continuously (Figure 10.30).

The key to stabilizing the photodiode amplifier is to introduce a pole so that the noise-gain curve flattens out before it hits the AOL curve. The rate of closure after introducing the pole is 20dB/decade. Figure 10.31 shows the stable example where noise gain intersects with AOL at 20dB/decade. Note that a small change in capacitance or AOL can cause the amplifier to become unstable. Therefore, it is generally advisable to move f_p to a lower frequency for good design margin.

Figure 10.32 gives the equations for selecting a feedback capacitance that will ensure stability. There are two formulas for C_f. The simplified formula assumes that C_{in} is much greater than C_f. The second equation gives a more exact value that is not dependent on this assumption. Note that these formulas compute the minimum capacitance required for stability. Increasing C_f beyond the minimum will ensure design margin.

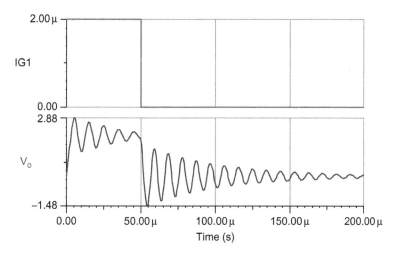

Figure 10.30: Step input shows stability issue

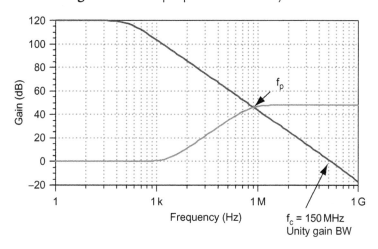

Figure 10.31: Pole from C_f stabilizes the circuit

$f_c = 150\,\text{MHz}$ Op-amp unity gain bandwidth

$C_{in} = 70\,\text{pF}$ Total input capacitance

$R_f = 1\,\text{M}\Omega$ Feedback resistance

$$C_f = \sqrt{\frac{C_{in}}{2\pi \cdot R_f \cdot f_c}} = 272.5\,\text{fF}$$ Simplified equation for minimum feedback capacitance assumes $C_{in} \gg C_f$

$$C_c = \frac{1}{2\pi \cdot R_f \cdot f_c}$$ Intermediate calculation used in more exact formula

$$C_{fe} = \frac{C_c}{2} \cdot \left(1 + \sqrt{1 + \frac{4C_{in}}{C_c}}\right) = 273.1\,\text{fF}$$ More exact formula for feedback capacitance

Figure 10.32: Equations for selecting minimum C_f for stability

Chapter Summary

- Photodiodes convert light to electrical current.
- The photodiode model contains shunt resistance, dark current, light current, and junction capacitance.
- Photodiode sensitivity to light is dependent on wavelength.
- Photodiode junction capacitance is dependent on the bias voltage.
- Signal bandwidth on a simple transimpedance amplifier is dependent on the feedback resistor and capacitor.
- The photodiode amplifier noise sources are photodiode current noise, op -amp current noise, op-amp voltage noise, and thermal noise from the feedback resistor.
- Analysis of the op-amp voltage noise requires breaking the spectral density curve into five different regions.
- Op-amp noise voltage gain peaking contributes significant noise.
- The amplitude of the noise voltage gain peak is dependent on the feedback capacitor.
- The photodiode amplifier has inherent stability issues. Choosing the appropriate value of feedback capacitance ensures stability.

Questions

10.1 What effect does increasing the reverse bias voltage have on junction photodiode junction capacitance?

10.2 What effect does decreasing photodiode junction capacitance have on:
  a. Noise peaking?
  b. Stability?
  c. Signal bandwidth?

10.3 Assuming that you have a fixed input capacitance. What circuit changes can be made to decrease the magnitude of the noise gain peak?

10.4 For the circuit below, use SPICE to graph Aol, noise gain, and signal gain. You will need to add some test components and to use the post processor.
  a. Is the circuit stable?
  b. What is the signal bandwidth?

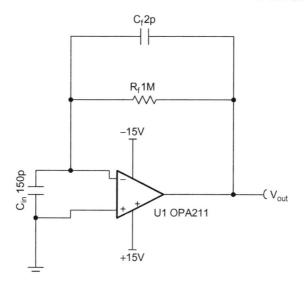

Further Reading

Graeme, J. Photodiode Amplifiers: OP AMP Solutions. McGraw-Hill Professional.
Green, T., 1996. Operational Amplifier Stability, <http://www.en-genius.net/>, McGraw-Hill.
    <http://www.hamamatsu.com/>, Photodiode Technical Information.

Photodiode Noise Amplifier Example Results

This chapter covers a sample calculation simulations and measurements of the photodiode circuit introduced in Chapter 10. Also included in Chapter 11 are real-world measurement results, and simulation results.

11.1 Photodiode Example Specifications

Table 11.1 shows some typical specifications for a photodiode. The photodiode used in this example is intended for lower frequency applications like smoke detectors, dimmers, and bar code scanners.

11.2 Photodiode Current Noise Calculations

Figure 11.1 calculates the total current noise spectral density from the photodiode. Part of this current noise is from the photodiode shunt resistance and part is from shot noise. One shot noise term is zero because the diode has no photocurrent.

11.3 Op-amp Specifications

Table 11.2 is a subset of the OPA827 data sheet. The key parameters are shown in bold.

11.4 Op-amp Voltage Noise Calculations

Figure 11.2 shows calculations of the key poles and zeros in the voltage noise-gain curve. The poles and zeros separate the different regions of the noise-gain curve. The poles and zeros depend on photodiode capacitance, feedback resistance R_f, feedback capacitance C_f, and the unity gain bandwidth of the op-amp used.

Table 11.1: Photodiode Specifications

Symbol	Characteristic	Condition	Min	Typ	Max	Units
R_{sh}	Shunt resistance	$V_r = 10\,mV$	100	150		$M\Omega$
C_j	Junction capacitance	$V_r = 0\,V$		70		pF
		$V_r = 10\,V$		10	25	
I_D	Dark current	$V_r = 0\,V$		2	30	nA
V_{BR}	Breakdown voltage	$I = 10\,\mu A$	30	75		V

$k_b := 1.38 \times 10^{-23}$ J/K Boltzmann constant

$T_n := 298$ K Temperature in Kelvin (289 K = 25°C)

$R_{sh} := 150 \times 10^6 \Omega$ Shunt resistance in photodiode

$i_j = \sqrt{\dfrac{4 \cdot k_b \cdot T_n}{R_{sh}}} = 10.472 \times 10^{-15} \dfrac{A}{\sqrt{Hz}}$ Thermal (Johnson noise)

$q := 1.602 \times 10^{-19}$ C One electron charge

$I_D := 2 \times 10^{-9}$ A Dark current in photodiode

$i_{sL} = \sqrt{2q \cdot I_D} = 25.314 \times 10^{-15} \dfrac{A}{\sqrt{Hz}}$ Shot noise (dark)

$I_L := 0$ A Photo current in photodiode (our measurements are dark)

$i_{sL} := \sqrt{2q \cdot I_L}$ Shot noise (white light)

$i_{n\_diode} = \sqrt{i_j^2 + i_{sD}^2 + i_{sL}^2} = 27.4 \times 10^{-15} \dfrac{A}{\sqrt{Hz}}$ Total diode current noise

Figure 11.1: Photodiode current noise spectral density

Table 11.2: OPA827 Data Sheet

Parameter		Standard Grade OPA827			Units
		Min	Typ	Max	
Noise					
Input voltage noise					
f = 0.1–10 Hz	e_n		250		nVpp
Input voltage noise density					
f = 1 kHz	e_n		4		nV/$\sqrt{Hz}$
f = 10 kHz	e_n		3.8		nV/$\sqrt{Hz}$
Input current noise density					
f = 1 kHz	i_n		2.2		fA/$\sqrt{Hz}$
Input impedance					
Differential			10^{13} || 9		
Common mode			10^{13} || 9		
Frequency response					
Gain-bandwidth product GBW			22		MHz

Figure 11.3 calculates the 1/f noise voltage corner frequency. The 1/f corner frequency separates the 1/f region from the broadband region on the op-amp voltage noise spectral density curve.

$R_f := 100\,k\Omega$ Feedback resistance

$C_f := 4\,pF$ Feedback capacitor

$C_j := 70\,pF$ Photodiode junction capacitance (from photodiode manufacturer)

$C_{opa} := 18\,pF$ Op-amp input capacitance (OPA827 data sheet)

$C_i := C_j + C_{opa}$ Total input capacitance

$f_c := 22\,MHz$ Unity gain bandwidth (OPA827 data sheet)

$$f_p = \frac{1}{2\pi\,R_f\cdot C_f} = 398\,Hz$$

$$f_z = \frac{1}{2\pi\,R_f\cdot(C_i + C_f)} = 17.3\,kHz$$

$$f_i = \frac{C_f}{C_i + C_f}\cdot f_c = 957\,kHz$$

Figure 11.2: Poles and zeros in the noise-gain curve

$e_{nif} := 3.8\,\dfrac{nV}{\sqrt{Hz}}$ Broadband noise spectral density (OPA827 data sheet)

$f_L := 0.1\,Hz$ Lower bound on frequency (1/f region) (arbitrary lower bound of frequency)

$e_{at\_f} := 60\,\dfrac{nV}{\sqrt{Hz}}$ Flicker noise measured at f_L (OPA827 data sheet noise curve)

$$e_{fnorm} = e_{at\_f}\cdot\sqrt{f_L} = 10\,nV$$

$$f_f = \frac{e_{fnorm}^2}{e_{nif}^2} = 24.9\,Hz$$

Figure 11.3: 1/f Noise corner

Figure 11.4 shows the total voltage noise calculation. Each term E_{noe1} through E_{noe5} corresponds to a different region in the noise output voltage curve. The total noise is calculated by summing the root sum of the squares of each region term. Note that Regions 3, 4, and 5 dominate the total noise. In this case, the 1/f noise performance is not a significant issue.

11.5 Thermal (Resistor) Noise Calculations

Figure 11.5 shows the calculation for thermal (resistor) noise from the feedback resistor. The feedback capacitor limits this noise by limiting the bandwidth (f_p).

$$E_{noe1} = \sqrt{e_{nif}^2 \cdot f_f \cdot \ln\left(\frac{f_f}{f_L}\right)} = 44.9 \, nV$$

$$E_{noe2} = \sqrt{e_{nif}^2 \cdot (f_z - f_f)} = 499 \, nV$$

$$E_{noe3} = \sqrt{\left(\frac{e_{nif}}{f_z}\right)^2 \cdot \frac{f_p^3 - f_z^3}{3}} = 31.8 \, \mu V$$

$$E_{noe4} = \sqrt{\left(e_{nif} \cdot \frac{C_i + C_f}{C_f}\right)^2 (f_i - f_p)} = 65.3 \, \mu V$$

$$E_{noe5} = \sqrt{\frac{(e_{nif} \cdot f_c)^2}{f_i}} = 85.4 \, \mu V$$

$$E_{noe} = \sqrt{E_{noe1}^2 + E_{noe2}^2 + E_{noe3}^2 + E_{noe4}^2 + E_{noe5}^2} = 112 \, \mu V$$

Figure 11.4: Total RMS output noise from op-amp voltage noise source

$R_f = 100 \, k\Omega$ Feedback resistance

$k_b = 1.38 \times 10^{-23}$ J/K   Boltzmann constant

$T_n = 298 \, K$ Temperature in Kelvin (25°C)

$f_p = 397.887 \times 10^3$ Hz   Transconductance bandwidth

$K_n = 1.57$ Noise current from OPA827 data sheet

$BW_n = K_n \cdot f_p$ Noise bandwidth (brick wall filter)

$e_{n\_r} = \sqrt{4k_b \cdot T_n \cdot R_f \cdot BW_n} = 33 \, \mu V$ Thermal noise at output

Figure 11.5: Total RMS output noise from feedback resistor thermal noise

11.6 Op-amp Current Noise Calculations

Figure 11.6 shows the hand calculation for the transimpedance amplifier current noise. The current noise is translated into a voltage noise by the feedback resistor. The feedback capacitor limits this noise by limiting the bandwidth (f_p).

11.7 Total Noise for Example Transimpedance Amplifier

Figure 11.7 is the total noise including all the noise components considered in this analysis (i.e., the op-amp voltage noise, resistor noise, and current noise). In this case, the op-amp voltage noise is the dominant noise source.

$R_f = 100 \times 10^3 \Omega$ Feedback resistance

$f_p = 397.887 \times 10^3 Hz$ Transconductance bandwidth

$i_{n\_opa} = 2.2 \times 10^{-15} \dfrac{A}{\sqrt{Hz}}$ Noise current from OPA827 data sheet

$i_{n\_diode} = 27.395 \times 10^{-15} \dfrac{A}{\sqrt{Hz}}$ Noise current from diode (calculated)

$i_{n\_total} = \sqrt{i_{n\_opa}^2 + i_{n\_diode}^2}$ Total noise current

$K_n = 1.57$ Noise bandwidth factor first-order filter

$BW_n = K_n \cdot f_p = 624\,kHz$ Noise bandwidth (brick wall filter)

$E_{noI} = i_{n\_total} \cdot R_f \cdot \sqrt{BW_n} = 2.17\,\mu V$ Current noise at output

Figure 11.6: Total RMS output noise from op-amp current noise

$E_{noe} = 112\,\mu V$ Op-amp voltage noise

$E_{noR} = 32\,\mu V$ Resistor noise

$E_{noI} = 2.17\,\mu V$ Op-amp current noise

$E_{no} = \sqrt{E_{noR}^2 + E_{noI}^2 + E_{noe}^2} = 117\,\mu V$ Total output noise for OPA827 transimpedance amp

Figure 11.7: Total RMS output noise including all components

11.8 Spice Analysis of Example Circuit

The circuit required to perform TINA Spice analysis for the example circuit is shown in Figure 11.9. Note that the full model for the diode is not required. In this analysis, we only use the diode junction capacitance $C_j = 70\,pF$.

Figure 11.8 shows the results of a TINA Spice noise simulation for the circuit shown in Figure 11.9. The output noise voltage spectral density curve shows noise peaking of $78\,nV/\sqrt{Hz}$. The noise density curve can be generated in TINA Spice using the "Output Noise" option on a "Noise Analysis." The total noise curve is the integration of the power spectral density. The total noise converges to 109-nV RMS; this corresponds well to the total noise of $117\,\mu V$ from hand calculations.

Figure 11.10 shows the effect of adjusting the feedback capacitor on bandwidth, spectral density, and total noise. The figure compares the 4-pF capacitor used in the previous calculations with a 2-pF capacitor. Generally, decreasing the feedback capacitor increases

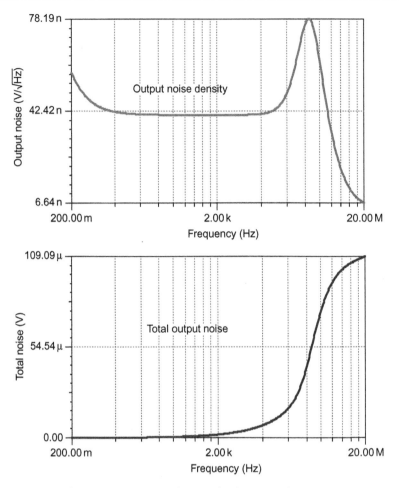

Figure 11.8: TINA Spice results for example circuit

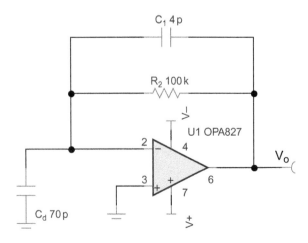

Figure 11.9: Example circuit for TINA Spice simulation

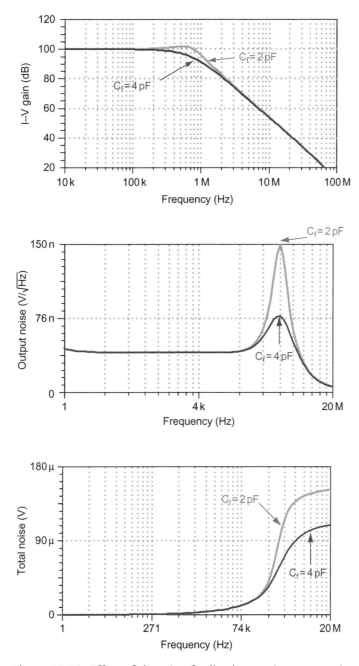

Figure 11.10: Effect of changing feedback capacitance on noise

the I–V bandwidth. In the example shown, the effect on the I–V bandwidth is not obvious, but there is noticeable gain peaking for the 2-pF case. The gain peaking is indicative of marginal stability. The noise spectral density curve shows significant noise peaking for the 2-pF capacitor. This additional noise peaking corresponds to attritional total noise.

11.9 Measuring the Noise for the Example Transimpedance Amplifier

Figure 11.11 shows the simplest approach to measuring noise. In this case, the amplifier output is directly connected to the measurement instrument. To determine if this approach can be used to successfully measure noise, you need to compare the expected noise level to the measurement noise floor.

Table 11.3 gives the specifications for two common examples of equipment used in noise measurements. The table also lists the noise signals from the transimpedance amplifier for comparison to the noise floor. The noise floor is close to the amplitude of the noise signal. Ideally, the noise floor would be significantly smaller than the noise signal. Without a postamplifier, the accuracy of the noise measurement would be marginal.

Figure 11.12 shows what can be used to improve the noise floor of the measurement circuit. The postamplifier increases the amplitude of the noise signal from the transimpedance amplifier so that the equipment is capable of measuring the noise. In order for this technique to be effective, the postamplifier must have an input noise that is significantly smaller than the output noise of the transimpedance amplifier. It is also important to make sure that the bandwidth of the postamplifier is wider than the amplifier under test. In cases where the amplifier under test has low noise, it can be challenging to select the postamplifier. For this example circuit, we will investigate using the OPA847 as a postamplifier.

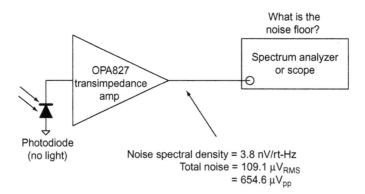

Figure 11.11: Checking equipment capability for measuring example circuit noise

Table 11.3: Noise Floor for Common-test Equipment

	Scope	Spectrum Analyzer	Noise Signal
Noise floor	48-μV RMS	10 nV/$\sqrt{\text{Hz}}$	109 μV RMS 3.8 nV/$\sqrt{\text{Hz}}$
Range	1 mV/dev		
Input impedance	1 MΩ	50 Ω	

Table 11.4 shows the effect of the gain on the noise signal from the transimpedance amplifier. The point is to ensure that the gain sufficiently increases the amplitude of the noise signal so that it can be easily measured by the test equipment. In Chapter 1, we have learned that noise signals three times greater than the noise floor dominate. In that example, the noise signal was amplified so that it was substantially larger than the noise floor to further minimize error. Note that most equipment has poor resolution near the noise floor, so operating significantly above the noise floor improves accuracy.

Figure 11.13 shows the bandwidth calculation for the postamplifier. Note that the postamplifier bandwidth must be greater than the transimpedance amplifier bandwidth for

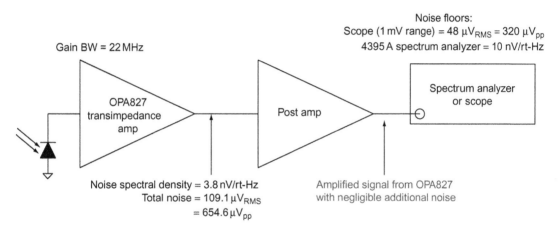

Figure 11.12: Low-noise postamplifier increases noise to a measurable level

Table 11.4: Amplify the Noise Signal to Improve Noise-measurement Accuracy

	Scope	Spectrum Analyzer	Noise Signal	Noise Signal × 150
Noise floor	48-μV RMS		109-μV RMS	16.35-mV RMS
		$10\,nV/\sqrt{Hz}$	$3.8\,nV/\sqrt{Hz}$	$570\,nV/\sqrt{Hz}$
Range	1 mV/dev			
Input impedance	1 MΩ	50 Ω		

$$\text{Gain\_Bandwidth\_Product} = 3900\,\text{MHz} \quad \text{For OPA847 gain} > 50$$

$$\text{Post\_Amp\_Bandwidth} = \frac{\text{Gain\_Bandwidth\_Product}}{\text{Gain}} = \frac{3900\,\text{MHz}}{150} = 26\,\text{MHz}$$

$$\text{Transimpedance\_BW} = 22\,\text{MHz} \quad \text{Unity gain bandwidth of the transimpedance amplifier is lower than the postamplifier bandwidth}$$

Figure 11.13: Noise floor for common test equipment

accurate noise measurements. However, using a bandwidth significantly wider than the amplifier under test adds unnecessary noise. In our example, the postamplifier bandwidth is relatively close to transimpedance amplifier bandwidth (i.e., postamplifier bandwidth 26 MHz and transimpedance amplifier bandwidth 22 MHz).

The OPA827 is a low-noise wide-bandwidth device. Figure 11.14 shows the proposed postamplifier. The gain of the amplifier is set to 150. Note that the parallel combination of the feedback network was intentionally selected as a low resistance so that it does not contribute significant thermal noise. The 50-Ω output resistor is for impedance matching to 50-Ω input impedance equipment. The output capacitor allows for AC coupling to the test equipment.

Figure 11.15 shows the total noise simulation results for the postamplifier. The data from this simulation is used in the Table 11.5 to show that the postamplifier noise is small compared with the transimpedance amplifier.

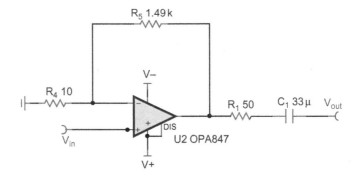

Figure 11.14: Low-noise wide-bandwidth postamplifier

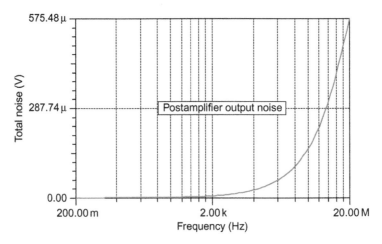

Figure 11.15: Postamplifier simulated noise

The noise output from the transimpedance amplifier adds with the input noise of the postamplifier. Table 11.5 illustrates that the noise referred-to-the–input (RTI) of the postamplifier is small compared with the noise at the output of the transimpedance amplifier. Thus, the postamplifier increases the amplitude of the noise from the transimpedance amplifier (150 ×), but does not add significant additional noise.

Figure 11.16 shows the hardware setup used for the transimpedance amplifier and the postamplifier. The printed circuit board on the right side is the photodiode amplifier. Note that the photodiode is covered with black tape to ensure that light does not affect the noise measurements.

Figure 11.17 shows the TINA Spice schematic for the photodiode amplifier and the postamplifier. Simulating this circuit allows us to confirm that the postamplifier increases the amplitude of the transimpedance amplifier noise, but does not contribute significant additional noise.

Table 11.5: Comparison of Postamplifier Input Noise with Transimpedance Output Noise

Transimpedance Amplifier Output Noise (From Simulation) (Figure 11.8)	Postamplifier Input Noise (RTI) RTO/150	Postamplifier Noise RTO (150 × RTI) (From Simulation) (Figure 11.13)
109-µV RMS 3.8 nV/√Hz	3.83-µV RMS 0.85 nV/√Hz	575.5-µV RMS 127.5-nV RMS

Figure 11.16: OPA847EVM used for postamplifier

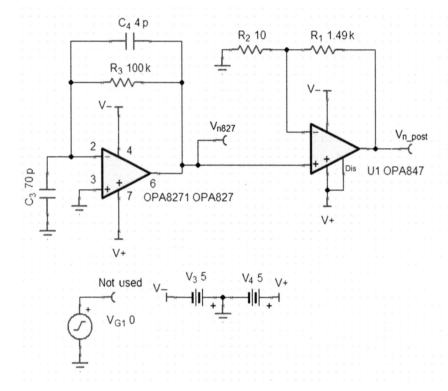

Figure 11.17: Transimpedance amplifier and postamplifier

Figure 11.18 shows the results of simulation of the circuit in Figure 11.17. The important point is that the postamplifier amplifies the noise by a factor of 150, but does not add significant noise. The results from the simulation are summarized in Table 11.6.

Figure 11.19 shows the connections to the spectrum analyzer and oscilloscope. Note that the spectrum analyzer has a 50-Ω input impedance. This is common on high-frequency test equipment. Note that the 50-Ω impedance creates a voltage divider with the output impedance of the postamplifier.

Figure 11.20 shows the oscilloscope noise measurement for the photodiode and postamplifier circuit.

Table 11.7 summarizes the results for the calculation, simulation, and measurement of this circuit. In this case, the measured results are about 30% larger than the calculated and simulated results. This difference could be the result of component tolerance or parasitic capacitance.

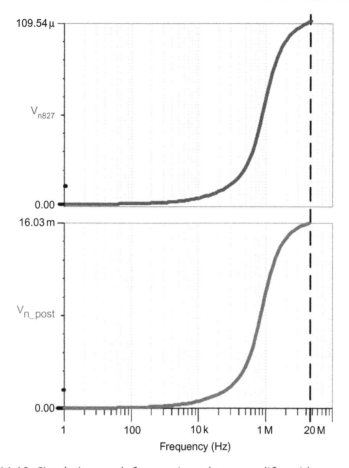

Figure 11.18: Simulation result for transimpedance amplifier with postamplifier

Table 11.6: Simulation Results Showing Insignificant Postamplifier Noise

V_n OPA827 Transimpedance Amplifier (Table 11.5)	V_n OPA847 Postamplifier (Table 11.5)	V_n Post/Gain 16mV/150
109.4-μV RMS	16-mV RMS	106.8-mV RMS

Figure 11.21 shows a comparison of the spectrum analyzer measurement and the simulated output voltage spectral density. Note that the spectral density curve is displayed with a linear frequency access. The TINA Spice spectral density results are shown with both the linear and the logarithmic axes.

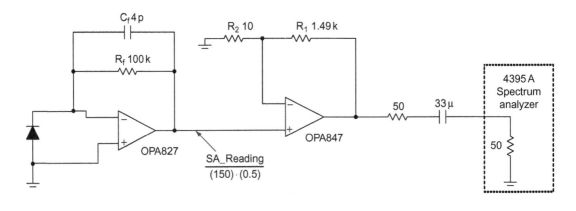

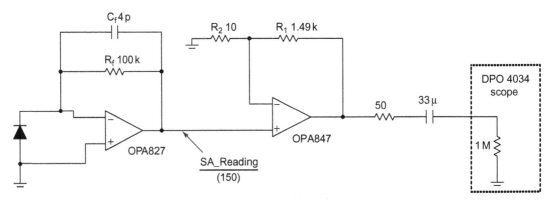

Figure 11.19: Connection to test equipment

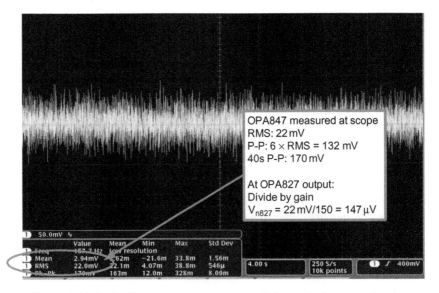

Figure 11.20: Oscilloscope measurement of photodiode noise circuit

Table 11.7: Summary of Calculated, Simulated, and Measured Results

Calculated (RMS)	Simulated (RMS)	Measured (RMS)
116.7 μV	109.1 μV	147 μV

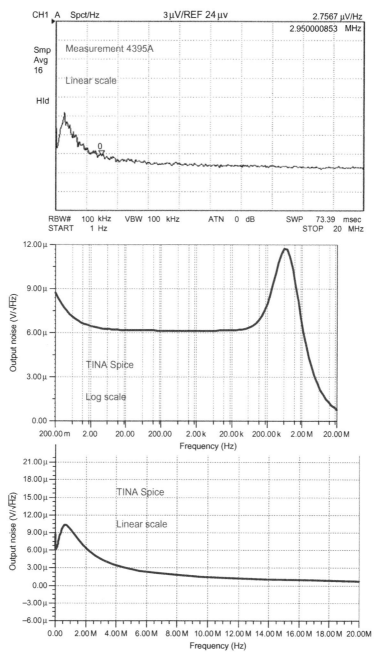

Figure 11.21: Spectral density measurement compared to simulation results

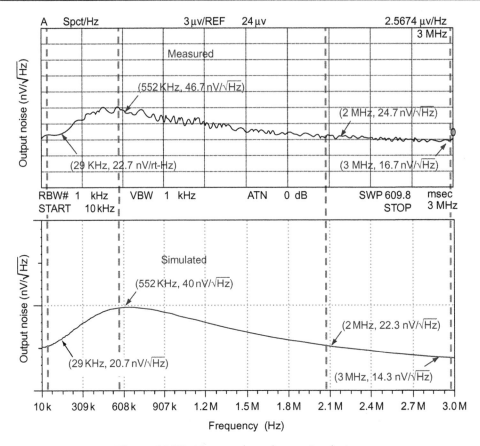

Figure 11.22: Measured results vs. simulations

Figure 11.22 shows the measured vs. simulated spectral density with key points on the curve identified. The measured spectral density curve compares well with the simulations. Any difference is likely the result of stray capacitance and component tolerance.

Chapter Summary

- Chapter 11 covers example calculations using the theory from Chapter 10.
  - OPA827 is used with a 70pF, 150 MΩ photodiode
- Chapter 11 also covers an example simulation of the circuit used in the example calculations
- Chapter 11 covers measurement of the example photodiode circuit
  - A post amplifier is used to amplify the noise so that the test equipment can measure the noise

- The post amplifier bandwidth must be wider then the noise signal being measured
- The post amplifier noise floor must be at least three time lower then noise being measured

Questions

11.1 For the diode with the characteristics below, what is the total current noise for the diode?

Symbol	Characteristic	Condition	Min	Typ	Max	Units
Rsh	Shunt Resistance	Vr = 10 mV	200	250		MΩ
Cj	Junction Capacitance	Vr = 0V		40		pF
		Vr = 10V		8	12	
I_D	Dark Current	Vr = 0V		10	80	nA
V_{BR}	Breakdown Voltage	I = 10 μA	20	55		V

11.2 For the circuit shown below, calculate the following?
   a. The key frequencies f_p, f_c, f_i, and f_f.
   b. Total RMS output noise from op-amp voltage noise source. E_{noe1}, E_{noe2}, E_{noe3}, E_{noe4}, E_{noe5}, E_{noe}.
   c. Total RMS output noise from current noise source. Assume the same diode is used from question 11.1.
   d. Total RMS output noise from all components.

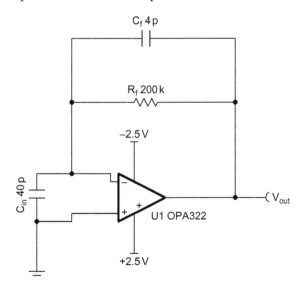

11.3 Simulate the total output noise and the noise spectral density circuit shown in 11.2.

11.4 Simulate the total output noise and the noise spectral density for the circuit shown below. Compare the spectral density curves from question 11.3 and 11.4. Why is the noise peak more noticeable in the circuit of 11.3?

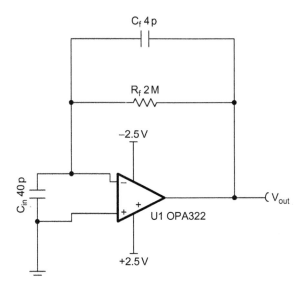

Further Reading

Graeme, J., 1996. Photodiode Amplifiers: OP AMP Solutions. McGraw-Hill, New York.

Green, T. Operational Amplifier Stability, <http://www.en-genius.net/> <http://www.hamamatsu.com/>, Photodiode Technical Information.

Glossary

1/f corner frequency The frequency where noise transitions from the 1/f region to the broadband region. See Figure G1

$$e_{fnorm} = e_{at\_f} \cdot \sqrt{f_L}$$ Noise normalized to 1 Hz

$$f_f = \frac{e_{fnorm}^2}{e_{nif}^2}$$ 1/f noise corner

where
e_{fnorm} is noise normalized to 1 Hz
e_{at_f} is noise at frequency f_L
e_{nif} is noise in broadband region
f_L is frequency in 1/f region
f_f is 1/f noise corner

Figure G1: Equation for calculating the 1/f noise corner

1/f Noise A form of intrinsic noise that has a power spectral density that is inversely proportional to frequency. For voltage or current spectral density, 1/f noise is inversely proportional to the square root of frequency. 1/f noise is typically an issue for low frequency applications (i.e., less than 1 kHz). See Figure G2

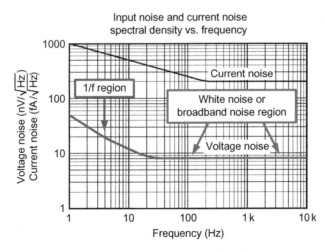

Figure G2: 1/f noise is inversely proportional to the square root of frequency

Auto-zero amplifier The auto-zero amplifier periodically samples the offset and stores it on a capacitor. The sampled offset is used to correct offset error and offset drift. This technique also eliminates flicker noise.
Averaging circuit A circuit that averages the signal at each of its multiple inputs. In the context of noise, an averaging circuit can be used to reduce noise. In noise averaging, the inputs to multiple amplifiers will be

connected in parallel and the outputs will be averaged. The total output noise is divided by the square root of the number of amplifiers. Thus, averaging four signals will reduce the noise by a factor of two.

Base current Current flowing into the base of a bipolar transistor.

Bench test A bench test is a test involving simple test equipment. Typically bench tests are used to characterize a small number of devices, so test time is not a major concern.

Bimodal distribution A bimodal distribution is the combination of two separate distributions. This distribution will have two peaks. Popcorn noise can have a bimodal or multimodal distribution. See Figure G3.

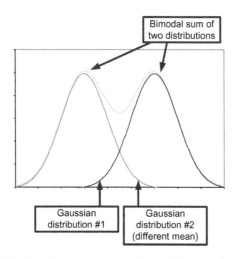

Figure G3: Two Gaussian curves form bimodal distribution

Bipolar transistor current gain The ratio of collector current to base current in a bipolar transistor is called current gain (β).

Bipolar transistors A three terminal semiconductor device that amplifies signals through current gain. Bipolar transistors are used internally in many op-amps. Bipolar amplifiers tend to have better voltage noise than CMOS amplifiers for a given quiescent current. CMOS amplifiers tend to have better current noise than bipolar amplifiers.

Brick wall factor (K_n) This factor can be used to convert an N-th order filter to a brick wall filter. For example, a first order filter can be converted to a brick wall filter by multiplying by the brick wall factor 1.57.

Brick wall filter A filter that has an infinitely steep roll-off. The pass band transits directly to the stop band so that the gain vs. frequency response looks like a brick wall. In noise calculations, the bandwidth of a brick wall filter can be used to represent the noise bandwidth of a system. In other words, the brick wall filter will have equivalent area under the power spectral density curve to the filter it represents. See Figure G4.

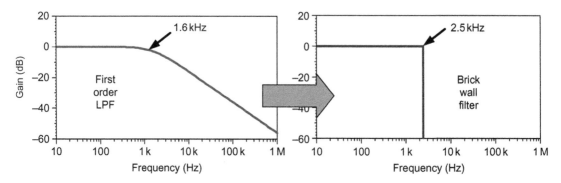

Figure G4: Brick wall equivalent to first order low pass filter

Broadband noise Broadband normal refers to a wide bandwidth signal range. In the case of noise, broadband is synonymous with white noise because higher frequency noise is normally in the flat spectral density region and outside of the 1/f region.

Burst noise See Popcorn noise.

Calibration feed through On zero drift amplifiers the calibration signal can generate noise. Typically this noise (and its harmonics) shows up as a spike in the spectral density curve.

Chopper-stabilized amplifier The chopper-stabilized amplifier polarity is synchronously flipped at the input and output. The net result is that signals passing through the amplifier are unaffected, but the offset polarity is flipped polarity every half cycle. Flipping the polarity of the offset effectively converts the offset to a square wave with an average value of zero. The square wave is filtered out with a synchronous filter. This technique also eliminates flicker noise

Closed loop gain The voltage gain of an op-amp circuit includes the effects of the feedback network. For example, closed loop gain for a noninverting amplifier is given by the equation Gain = Rf/R1 + 1. Closed loop gain is normally significantly smaller than the open loop gain.

Common mode signal The average signal applied to both the inputs of the differential amplifiers.
$V_{cm} = (V_{in1} + V_{in2})/2$

Correlated noise If two noise signals are mathematically related to each other they are said to be correlated.

Decibel milliwatts (dBm) Power measured with reference to 1 mW. With a spectrum analyzer, this power is measured across the 50-Ω input impedance.

Differential amplifier A differential amplifier is designed to amplify differential signals and reject common mode signals. The input impedance of differential amplifiers is determined by the resistor networks used to set the gain. Normally the input impedance of a differential amplifier is considerably lower than the input impedance of an instrumentation amplifier. See Figure G5.

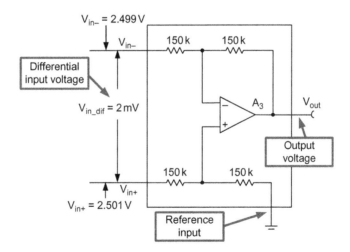

Figure G5: Differential amplifier schematic

Differential signal The signal measured between both inputs. $V_{dif} = V_{in1} - V_{in2}$

Dominant pole On an op-amp open loop gain curve the dominant pole is the frequency where the AOL curve begins to decrease with frequency. See Figure G6.

Electromagnetic interference (EMI) Electromagnetic interference (EMI) includes all sorts of extrinsic noise (e.g., motors, 60-Hz power, digital switching, cell phone, etc.). EMI can be coupled into an amplifier circuit through radiated or conducted emissions. See Radio frequency interference (RFI).

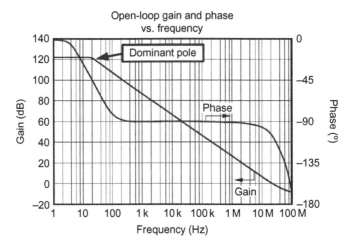

Figure G6: Dominant pole location on Aol Curve

Extrinsic noise Noise generated outside the device. This type of noise is coupled into sensitive circuits through electromagnetic, inductive, and capacitive coupling. Common examples of extrinsic noise are 60-Hz powered line noise.

Feedback capacitor A feedback capacitor can limit bandwidth and consequently limit the total noise. At high frequency, the feedback capacitor will effectively short out the feedback resistor. Shorting the feedback resistor effectively reduces the noise gain to unity. This method is most effective for high gain amplifiers.

Field effect transistors (FET) A three terminal semiconductor device that amplifies signals through voltage gain. MOS transistors are used internally in many op-amps. Bipolar amplifiers tend to have better voltage noise than MOS amplifiers for a given quiescent current. MOS amplifiers tend to have better current noise than bipolar amplifiers.

Flicker noise See 1/f noise.

Frequency domain The frequency domain is used to measure the frequency content of signals. In the case of noise, the frequency domain is shown in a spectral density plot. The spectrum analyzer is an example of an instrument that measures spectral density.

Gain bandwidth product This specification is given in op-amp data sheets and is used to compute the closed loop gain of an op-amp circuit. Closed_Loop_Bandwidth = Gain_Bandwidth_Product / Closed_Loop_Gain

Gaussian distribution A bell shaped distribution often used to predict the probability that an event will occur in a given interval. Most random noise can be statistically represented using a Gaussian curve.

Input stage noise With instrumentation amplifiers or other multiple stage amplifiers there is noise associated with each stage. The input stage noise is the noise associated with the first stage. If the first amplifier stage is in gain, it is normally the dominant source of noise.

Instrumentation amplifier An instrument amplifier is designed to amplify differential signals and reject common mode signals. Instrument amplifiers have high impedance inputs. See Figure G7.

Integrated noise See Total noise.

Intrinsic noise Noise that is generated by the device itself. A common example of intrinsic noise is thermal noise. In thermal noise, the noise is generated by random motion of electrons inside the device. Another common example of intrinsic noise is shot noise.

Junction capacitance An internal parasitic capacitance in parallel with the photodiode. The junction capacitance is affected by reverse voltage applied to the photodiode.

Lot to lot variation Each semiconductor lot that is fabricated can have different characteristics from other lots. Ideally the differences are small. In practice, some lots may exhibit shifts in some parameters. For example, one particular lot may have high current noise. Another lot may have issues with popcorn noise.

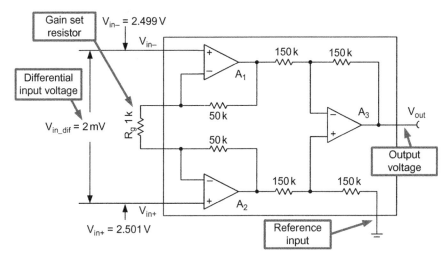

Figure G7: Instrumentation amplifier schematic

Low frequency noise Op-amp data sheets often have a 0.1–10 Hz peak-to-peak noise specification. This is a measurement of the low-frequency noise. Typically this is in the 1/f noise region.

Mean For a discrete set of data points the mean is the sum of each value divided by the number of values in the population. The arithmetic average of the data. See Figure G8.

Mean defined for a discrete statistical population

$$\mu = \frac{1}{b-a} \int_a^b g(t)\, dt \qquad \text{Continuous form}$$

$$\mu = \frac{1}{n} \sum_{i=1}^{n} x_i \qquad \text{Discrete form}$$

Figure G8: Equations for Mean

Nodal analysis Analysis technique that develops equation by considering that the sum of currents at a given node is zero. This technique can be useful for analyzing complex op-amp circuits.

Noise An error signal that is not desired. Noise generated by the circuit components itself (intrinsic noise) typically appears as a random change in voltage or current.

Noise bandwidth An AC transfer function shows gain vs. frequency. The bandwidth of an AC transfer function can be limited by various types of filters. The noise bandwidth is a theoretical rectangular filter (brick wall filter) that is equivalent to the actual physical filter. See Figure G9.

Noise floor The noise measured by an instrument without any external circuit connected. It is not possible for an instrument to measure noise levels below the noise floor.

Noise gain The gain seen by the voltage noise source. Noise gain can be different than signal gain. For example, the inverting amplifier signal gain is different than the noise gain because the noise signal is at the noninverting input.

$$BW_n = \frac{1}{\left(\left|A_{vo}\right|\right)^2} \cdot \int_0^{\infty} \left(\left|A_v(f)\right|\right)^2 df$$

where
BW$_n$ is noise bandwidth
$A_v(f)$ is AC transfer characteristic vs. frequency
A_{vo} is maximum magnitude of transfer function
f   is frequency

Figure G9: Mathematical definition of noise bandwidth

Noise gain peaking In transimpedance amplifier, the input capacitance of an amplifier causes noise gain to increase or peak. This peak can contribute significant total noise.

Noise model The noise model of a component shows the noise sources that are generated by the device. For example, an op-amp noise model has a noise current source and a noise voltage source. See Figure G10.

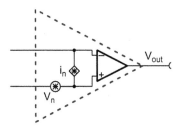

Figure G10: Noise model for op-amp

Non-Gaussian A Gaussian distribution has a bell shape and follows the mathematical function given below. Any distribution that does not follow this mathematical function is non-Gaussian. Popcorn noise is an example of a signal that has a non-Gaussian distribution.

Offset temperature drift Op-amp input offset voltage will change or drift with temperature. This drift is minimized in precision low-drift amplifiers by trimming during manufacturing. Zero-drift amplifiers minimize offset drift with a digital offset calibration circuit. Temperature offset drift can often be mistaken for low frequency noise (1/f noise).

Op-amp stability A stable op-amp can properly amplify signals inside its defined bandwidth. Unstable amplifiers have substantial overshoot, ringing, or sustained oscillations. Stability is most often a concern when reactive loads are connected to amplifiers. The input capacitance of a standard transimpedance amplifier can cause instability if the amplifier is not properly stabilized.

Open loop gain This is the gain of an op-amp without feedback. Normally high open loop gain is desirable. In noise analysis the open loop gain and the gain bandwidth product are used to determine the dominant pole.

Oscilloscope An instrument that can be used to measure noise in the time domain.

Output stage noise With instrumentation amplifiers or other multiple stage amplifiers there is noise associated with each stage. The output stage noise is the noise associated with the final stage. The output stage noise is most important when the input stage is in low gain. When the input stage is in high gain, the output stage noise can often be neglected.

Peak-to-peak The maximum deviation of an AC signal. The difference between the highest value and the lowest value on the signal. See Figure G11.

Photodiode A semiconductor diode that is designed to convert light into current.

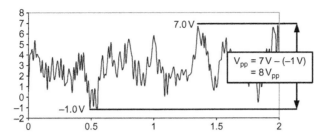

Figure G11: Peak-to-peak

Photodiode amplifier Photodiodes have a photovoltaic and photoconductive mode of operation. In the photovoltaic mode of operation the voltage across the photodiode is amplified. In the photoconductive mode the photodiode current is converted to voltage with a transimpedance amplifier. In this text, we focus on the photoconductive mode.

Photodiode current noise This is the combined noise current from the diode shot noise and the shunt resistance thermal noise.

Photodiode dark current The current that flows in the photodiode without applied light.

Photodiode shunt resistance An internal parasitic resistance in parallel with the photodiode.

Popcorn noise Noise that makes discrete jumps or steps from one value to another. Popcorn noise is caused by semiconductor defects or processing issues. A device with popcorn noise is considered to be defective.

Postamplifier A postamplifier can be used to amplify the noise of a circuit so that it can be measured by standard test equipment. This circuit is used in cases where the noise floor of the test equipment is not sufficient.

Probability density function The statistical function that defines the probability of a random variable. For a Gaussian distribution, it is the function that describes the shape of the curve. The integral of this function is the probability distribution function. See Figure G12.

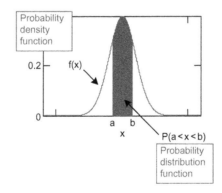

(1.2) Probability density function for normal (Gaussian) distribution

$$f(x) = \frac{1}{\sigma \cdot \sqrt{2\pi}} \cdot e^{[-(x-\mu)^2/2\sigma^2]}$$

where
f(x) is the probability that x will be measured at any instant in time
x is the random variable. In this case
μ is the mean value
σ is the standard deviation

Figure G12: Probability density function

Probability distribution function The integral of the probability density function is called the probability distribution function. This function gives the probability that an event will occur over a given interval. For example if $P(-1 < X < +1) = 0.1$, then there is a 10% chance that X is in the interval $-1 < X < +1$. In noise calculations, the probability distribution function of a Gaussian curve is used to estimate the peak-to-peak noise based on the standard deviation. To estimate peak-to-peak consider that 99.5% of noise is inside of ± 3 standard deviations. See Figure G13.

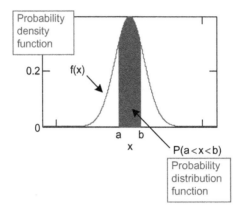

$$P(a<x<b) = \int_a^b f(x)\, dx = \int_a^b \frac{1}{\sigma \cdot \sqrt{2\pi}} \cdot e^{[-(x-\mu)^2/2\sigma^2]}\, dx$$

(1.3) Probability distribution function for normal (Gaussian) distribution

where
$P(a<x<b)$ is the probability that x will be in the interval (a, b) in any instant in time. For example, $P(-1 < x < +1) = 0.3$ means that there is a 30% chance that x will be between -1 and 1 for any measurement.
x is the random variable. In this case noise voltage.
μ is the mean value
σ is the standard deviation

Figure G13: Probability distribution function

Production test A production test utilizes automatic test equipment (ATE). Typical production tests are set up to test thousands or millions of devices. Most op-amp parameters are verified using a production test for each device that is manufactured. Test time is an important concern in production testing because the test adds additional cost to the device.

Proportionate to absolute temperature (PTAT) In a PTAT transistor amplifier bias, the collector (drain) current increases directly proportional to temperature. The type of temperature control on bias has some affect on the noise vs. temperature.

Quiescent current The minimum operating current of an op-amp. This specification is listed in the op-amp data sheet. Low noise op-amps typically draw higher quiescent current (I_q).

Radio frequency interference (RFI) Radio frequency interference (RFI) is a signal from a transmitter (e.g., cell phone, television station, etc.) that will induce noise in an amplifier circuit. In this case, the printed circuit traces and component leads act as unintended antennas that pick up these signals. RFI is a specific type of electromagnetic interference (EMI). See Electromagnetic interference (EMI).

Random telegraph signals (RTS) See Popcorn noise.

Reference buffer In a single supply instrumentation amplifier circuit, the reference input is normally connected to one half the power supply voltage. This voltage is often developed using a voltage divider but must be buffered to prevent loading.

Resolution bandwidth In spectrum analyzers, this is the bandwidth of a bandpass filter that is swept across frequency to measure the magnitude of signals at different frequencies. Using a narrow resolution bandwidth produces more accurate measurements but increases the measurement time.

Root mean square (RMS) For AC signals, the RMS is the effective equivalent value that delivers the same power to a load as a DC signal. For example, a 1-V AC sinusoidal waveform has an RMS of 0.707 V. This means that a DC voltage source of 0.707 V would produce the same power (self heating) as a 1-V peak AC sinusoidal wave. Noise is typically represented as an RMS value. See Figure G14.

Root mean square (RMS) defined for a probability distribution function
This is the same as σ if $\mu = 0$

$$RMS = \sqrt{\frac{1}{b-a}\left(\int_{a}^{b} g(t)^2 \, dt\right)} \qquad \text{Continuous form}$$

$$RMS = \sqrt{\frac{1}{n}\sum_{i=1}^{n} x_i^2} \qquad \text{Discrete form}$$

Figure G14: Equations for RMS

Semiconductor defects A semiconductor defect is caused by contamination or misprocessing of the fabrication. Semiconductor defects typically cause failure of operation or seriously degrade performance.

Semiconductor process Op-amps are semiconductor devices. The process of building the semiconductor devices involves many chemical and photolithography steps. The key thing to understand is that variations in the semiconductor process will translate into variations in op-amp performance. Some op-amp specifications do not change much with process variation and others are strongly affected. For example, broadband noise spectral density is not strongly affected by process variations (i.e., $\pm 10\%$), but flicker is strongly affected by process variations ($\pm 3 \times$ for some processes). In addition to normal fluctuations on process variation, it is possible to introduce defects by improperly processing semiconductors. Defects can cause popcorn noise and other abnormal behavior.

Shielding Placing an electrical circuit in a metal enclosure to prevent the pickup of extrinsic noise signals is called shielding.

Spectral density Noise per unit frequency. Power spectral density is measured in W/Hz. Voltage spectral density is measured in units of V/rt-Hz.

Spectrum analyzer An instrument that can be used to measure the spectral density of noise.

Spice (Simulation Program with Integrated Circuit Emphasis) A software tool that enables simulation of electronic circuits. Spice enables AC, DC, transient, and noise simulations. In noise analysis, Spice can be used to predict the total noise output for a circuit.

Standard deviation The standard deviation is a measure of the variability of a statistical population. For noise six times the standard deviation can be used to estimate the peak-to-peak value of a signal. See Figure G15.

Test limits Maximum and minimum values used to define if a test passes or fails.

Thermal noise Thermal noise is noise generated by random motion of electrons inside a resistance. Thermal noise has a flat spectral density curve; i.e., the noise power is equivalent over frequency. See Figure G16.

Time domain The time domain is used to show how a quantity (i.e., voltage or current) changes with time. The oscilloscope is an example of an instrument that creates a time domain graph of voltage vs. time.

Standard deviation defined for a probability distribution function

$$\sigma = \sqrt{\frac{1}{b-a} \int_{a}^{b} (g(t) - \mu)^2 \, dt} \qquad \text{Continuous form}$$

$$\sigma = \sqrt{\sigma^2} = \sqrt{\frac{1}{n} \sum_{i=1}^{n} (x_i - \mu)^2} \qquad \text{Discrete form}$$

Figure G15: Equations for standard deviation

$$e_n = \sqrt{4 \, k \, T \, R \, \Delta f} \qquad \text{(1.1) Thermal noise equation}$$

where
e_n is the RMS noise voltage
T is temperature in Kelvin (K)
R is resistance in Ohms (Ω)
Δf is noise bandwidth in Hertz (Hz)
k is Boltzmann's constant 1.381×10^{-23} J/K

Note: To convert degrees Celsius to Kelvin

$$T_k = 273.15 \, ^\circ C + T_c$$

Figure G16: Equation for thermal noise voltage

TINA Spice TINA Spice is a version of Spice developed by Design Soft. A free version of TINA Spice can be downloaded from Texas Instruments. A full featured version of TINA Spice can be purchased at www.designsoftware.com

Total noise The RMS noise calculated by integrating the power spectral density over the noise bandwidth. For op-amps, the total noise includes the combined effects of nose voltage sources, noise current sources, and resistor noise sources. Figure G17.

$$\text{Noise power} = \int_{f_1}^{f_2} e_n^2 \, df \qquad \text{(1.10) RMS noise power over a frequency range.}$$
$$e_n \text{ is the noise voltage spectral density.}$$

$$\text{Noise voltage} = \sqrt{\int_{f_1}^{f_2} e_n^2 \, df} \qquad \text{(1.11) RMS noise voltage over a frequency range.}$$
$$e_n \text{ is the noise voltage spectral density.}$$

Figure G17: Equations for total RMS noise power and voltage

Transimpedance amplifier An amplifier that converts current to voltage. The gain of a transimpedance amplifier is given in volts per ampere (V/A).

True RMS DMM A digital multimeter that measures (calculates) the RMS signal level for an arbitrary wave shape. Some instruments can measure the RMS level for a sinusoidal wave but not for other wave shapes. A true RMS meter will properly measure RMS for any waveform type.

Uncorrelated noise When two noise sources are random and there is no mathematical relationship between the two sources they are called uncorrelated.

White noise White light is light with equal amounts of all colors. White noise has equal power spectral density over all frequencies (i.e., it has a flat spectral density curve). See Figure G18.

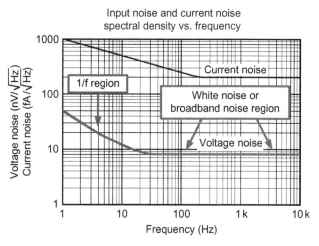

Figure G18: White noise has flat spectral density

Zero-drift amplifiers Amplifiers that use an internal digital circuit to correct input offset voltage and offset drift are called zero-drift amplifiers. These amplifiers do not have 1/f voltage noise. Two common zero-drift topologies are auto-zero and chopper-stabilized amplifier.

Zero-TC In a Zero-TC transistor amplifier bias, the collector (drain) current does not change with temperature.

Answers to Questions

Chapter 1

1.1 $E_n = \sqrt{4kTR\,\Delta f} = \sqrt{4(1.381 \times 10^{-23}\,\text{J/K})(298.15\,\text{K})(10\,\text{k}\Omega)(10\,\text{kHz})}$
                $= 1.283\text{-}\mu\text{V RMS}$

1.2 $E_{n\_pp} = 6\sigma = 6(1.283\mu\text{V}) = 7.698\mu\text{V}_{pp}$

1.3 $E_n = \sqrt{4kTR} = \sqrt{4(1.381 \times 10^{-23}\,\text{J/K})(298.15\,\text{K})(10\,\text{k}\Omega)} = 128.3\,(\text{nV}/\sqrt{\text{Hz}})$

1.4 $E_{nT} = \sqrt{E_{n1}^2 + E_{n2}^2} = \sqrt{(10\text{mV})^2 + (5\text{mV})^2} = 11.18\text{-mV RMS}$

1.5 $E_{nT} = \sqrt{E_{n1}^2 + E_{n2}^2 + 2CE_{n1}E_{n2}} = \sqrt{(10\text{mV})^2 + (5\text{mV})^2 + 2(-1)(10\text{mV})(5\text{mV})}$
                           $= 5\text{-mV RMS}$

1.6 $E_{RMS} = \dfrac{E_{npp}}{6} = \dfrac{25\text{mV}_{pp}}{6} = 4.167\text{-mV RMS}$

Chapter 2

2.1 $BW_n = f_H K_n = (1\text{kHz})(1.57) = 1.57\,\text{kHz}$

2.2

     a. $e_{fnorm} = e_{at\_f}\sqrt{f} = (60\,\text{nV}/\sqrt{\text{Hz}})\sqrt{0.1\text{Hz}} = 18.97\,\text{nV}$

        $E_{nf} = e_{fnorm}\sqrt{\ln\left(\dfrac{f_H}{f_L}\right)} = (18.97\,\text{nV})\sqrt{\ln\left(\dfrac{1\text{kHz}}{0.1\text{Hz}}\right)} = 57.6\text{-nV RMS}$

     b. $BW_n = f_H K_n = (1\text{kHz})(1.57) = 1.57\,\text{kHz}$

        $E_{nBB} = e_{BB}\sqrt{BW_n} = (3.8\,\text{nV}/\sqrt{\text{Hz}})\sqrt{1.57\text{kHz}} = 150.6\text{-nV RMS}$

2.3

a. $R_{eq} = \dfrac{R_f R_1}{R_f + R_1} = \dfrac{(1\,M\Omega)(100\,k\Omega)}{1\,M\Omega + 100\,k\Omega} = 90.8\,k\Omega$

$E_{n\_r} = \sqrt{4kTR\,\Delta f} = \sqrt{4(1.381 \times 10^{-23}\,J/K)(298.15\,K)(90.8\,k\Omega)(5\,kHz)}$
$= 2.734\text{-}\mu V\ RMS$

b. $I_{nBB} = i_{nBB}\sqrt{BW_n} = (5\,fA/\sqrt{Hz})\sqrt{(5\,kHz)} = 354\text{-}fA\ RMS$
$I_n = I_{nBB}$ Because no flicker component is specified
$E_{n\_i} = I_n R_{eq} = (354\,fA)(90.8\,k\Omega) = 32.1\text{-}nV\ RMS$

2.4 $E_{n\_in} = \sqrt{E_{n\_i}^2 + E_{n\_v}^2 + E_{n\_r}^2} = \sqrt{(100\,\mu V)^2 + (150\,\mu V)^2 + (75\,\mu V)^2}$
$= 195\text{-}\mu V\ RMS$

Chapter 3

3.1

a. $Closed\_Loop\_Bandwidth = \dfrac{Gain\_Bandwidth\_Product}{Noise\_Gain} = \dfrac{350\,kHz}{101} = 3.465\,kHz$

$BW_n = f_H K_n = (3.465\,kHz)(1.57) = 5.44\,kHz$

b. $E_{nBB} = e_{BB}\sqrt{BW_n} = 55\,nV/\sqrt{Hz}\sqrt{(5.44\,kHz)} = 4.057\text{-}\mu V\ RMS$

c. $I_{nBB} = i_{nBB}\sqrt{BW_n} = (100\,fA/\sqrt{Hz})\sqrt{(5.44\,kHz)} = 7.38\text{-}pA\ RMS$

$R_{eq1} = \dfrac{R_f R_1}{R_f + R_1} = \dfrac{(100\,k\Omega)(1\,k\Omega)}{100\,k\Omega + 1\,k\Omega} = 0.99\,k\Omega$

$E_{n\_i1} = I_n R_{eq} = (7.38\,pA)(0.99\,k\Omega) = 7.31\text{-}nV\ RMS$

$R_{eq2} = \dfrac{R_f R_1}{R_f + R_1} = \dfrac{(100\,k\Omega)(100\,k\Omega)}{100\,k\Omega + 100\,k\Omega} = 50\,k\Omega$

$E_{n\_i2} = I_n R_{eq} = (7.38\,pA)(50\,k\Omega) = 369\text{-}nV\ RMS$

$E_{n\_i} = \sqrt{E_{n\_i1}^2 + E_{n\_i2}^2} = \sqrt{(7.31\,nV)^2 + (369\,nV)^2} = 369.1\,nV$

d. $E_{n\_r1} = \sqrt{4kTR\,\Delta f} = \sqrt{4(1.381 \times 10^{-23}\,\text{J/K})(298.15\,\text{K})(0.99\,\text{k}\Omega)(5.44\,\text{kHz})}$
$$= 298\text{-nV RMS}$$

$E_{n\_r2} = \sqrt{4kTR\,\Delta f} = \sqrt{4(1.381 \times 10^{-23}\,\text{J/K})(298.15\,\text{K})(50\,\text{k}\Omega)(5.44\,\text{kHz})}$
$$= 2.117\text{-}\mu\text{V RMS}$$

$E_{n\_i} = \sqrt{E_{n\_r1}^2 + E_{n\_r2}^2} = \sqrt{(298\,\text{nV})^2 + (2.117\,\mu\text{V})^2} = 2.138\text{-}\mu\text{V RMS}$

e. $E_{n\_in} = \sqrt{E_{n\_i}^2 + E_{n\_v}^2 + E_{n\_r}^2} = \sqrt{(369.1\,\text{nV})^2 + (4.057\,\mu\text{V})^2 + (2.138\,\mu\text{V})^2}$
$$= 4.601\text{-}\mu\text{V RMS}$$

$E_{n\_out} = \text{Noise\_Gain} \cdot E_{n\_in} = 101(4.601\,\mu\text{V}) = 464\text{-}\mu\text{V RMS RTO}$

3.2

a. $\text{Closed\_Loop\_Bandwidth} = \dfrac{\text{Gain\_Bandwidth\_Product}}{\text{Noise\_Gain}} = \dfrac{18\,\text{MHz}}{11}$
$$= 1.636\,\text{MHz}$$

$BW_n = f_H K_n = (1.636\,\text{MHz})(1.57) = 2.569\,\text{MHz}$

b. $E_{nBB} = e_{BB}\sqrt{BW_n} = 2.2\,\text{nV}/\sqrt{\text{Hz}}\sqrt{(2.569\,\text{MHz})} = 3.526\text{-}\mu\text{V RMS}$
$e_{fnorm} = e_{at\_f}\sqrt{f} = 20\,\text{nV}/\sqrt{\text{Hz}}\sqrt{0.1\,\text{Hz}} = 6.325\,\text{nV}$

$E_{nf} = e_{fnorm}\sqrt{\ln\left(\dfrac{f_H}{f_L}\right)} = (6.325\,\text{nV})\sqrt{\ln\left(\dfrac{2.569\,\text{MHz}}{0.1\,\text{Hz}}\right)} = 24.3\text{-nV RMS}$

$E_{n\_v} = \sqrt{E_{nf}^2 + E_{nBB}^2} = \sqrt{(24.3\,\text{nV})^2 + (3.526\,\mu\text{V})^2} = 3.526\text{-}\mu\text{V RMS}$

c. $I_{nBB} = i_{nBB}\sqrt{BW_n} = (500\,\text{fA}/\sqrt{\text{Hz}})\sqrt{(2.569\,\text{MHz})} = 801.4\text{-pA RMS}$
$I_{fnorm} = I_{at\_f}\sqrt{f} = 3.75\,\text{pA}/\sqrt{\text{Hz}}\sqrt{0.1\,\text{Hz}} = 1.186\,\text{pA}$

$I_{nf} = I_{fnorm}\sqrt{\ln\left(\dfrac{f_H}{f_L}\right)} = (1.186\,\text{pA})\sqrt{\ln\left(\dfrac{2.569\,\text{MHz}}{0.1\,\text{Hz}}\right)} = 4.9\text{-pA RMS}$

$I_n = \sqrt{I_{nf}^2 + I_{nBB}^2} = \sqrt{(4.9\,\text{pA})^2 + (801.4\,\text{pA})^2} = 801.4\text{-pA RMS}$

$R_{eq1} = \dfrac{R_f R_1}{R_f + R_1} = \dfrac{(10\,\text{k}\Omega)(1\,\text{k}\Omega)}{10\,\text{k}\Omega + 1\,\text{k}\Omega} = 0.909\,\text{k}\Omega$

$E_{n\_i1} = I_n R_{eq} = (801.4\,\text{pA})(0.99\,\text{k}\Omega) = 7.31\text{-nV RMS}$

$E_{n\_i2} = I_n R_{in} = (801.4\,\text{pA})(100\,\text{k}\Omega) = 80.1\text{-}\mu\text{V RMS}$

$E_{n\_i} = \sqrt{E_{n\_i1}^2 + E_{n\_i2}^2} = \sqrt{(7.31\,\text{nV})^2 + (80.1\,\mu\text{V})^2} = 80.1\text{-}\mu\text{V RMS}$

d. $E_{n\_r1} = \sqrt{4kTR\Delta f}$

$$= \sqrt{4(1.381 \times 10^{-23}\,\text{J/K})(298.15\,\text{K})(0.909\,\text{k}\Omega)(5.44\,\text{kHz})} = 285\text{-nV RMS}$$

$E_{n\_r2} = \sqrt{4kTR\Delta f} = \sqrt{4(1.381 \times 10^{-23}\,\text{J/K})(298.15\,\text{K})(100\,\text{k}\Omega)(5.44\,\text{kHz})}$
$\qquad\qquad = 2.99\text{-}\mu\text{V RMS}$

$E_{n\_r} = \sqrt{E_{n\_r1}^2 + E_{n\_r2}^2} = \sqrt{(285\text{nV})^2 + (2.99\mu\text{V})^2} = 3\text{-}\mu\text{V RMS}$

e. $E_{n\_in} = \sqrt{E_{n\_i}^2 + E_{n\_v}^2 + E_{n\_r}^2} = \sqrt{(80.1\mu\text{V})^2 + (3.526\mu\text{V})^2 + (3\mu\text{V})^2}$
$\qquad\qquad = 80.2\text{-}\mu\text{V RMS}$

$E_{n\_out} = \text{Noise}\_\text{Gain} \cdot E_{n\_in} = 11(80.2\mu\text{V}) = 882\text{-}\mu\text{V RMS RTO}$

Chapter 4

4.1

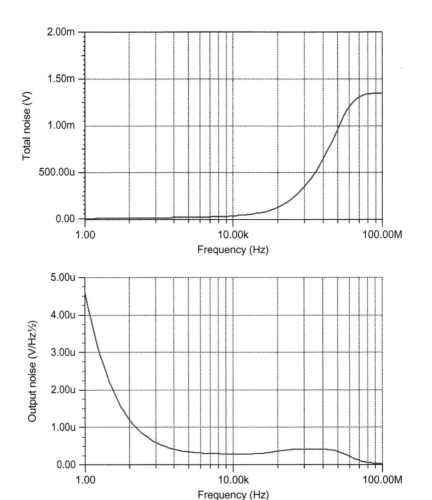

4.2 $\text{OLG} = 10^{132/20} = 3.981 \times 10^6$

$$\text{Dominant\_Pole} = \frac{\text{GBW}}{\text{OLG}} = \frac{18\,\text{MHz}}{3.981 \times 10^6} = 4.52\,\text{Hz}$$

4.3   Current source:

```
* BEGIN PROG NSE FEMTO AMP/RT-HZ
.SUBCKT FEMT 1 2
* BEGIN SETUP OF NOISE GEN - FEMPTOAMPS/RT-HZ
* INPUT THREE VARIABLES
* SET UP INSE 1/F
* FA/RHZ AT 1/F FREQ
.PARAM NLFF = 2.2
* FREQ FOR 1/F VAL
.PARAM FLWF = 0.001
* SET UP INSE FB
* FA/RHZ FLATBAND
.PARAM NVRF = 2.2
* END USER INPUT
* START CALC VALS
.PARAM GLFF = {PWR(FLWF,0.25)*NLFF/1164}
.PARAM RNVF = {1.184*PWR(NVRF,2)}
.MODEL DVNF D KF = {PWR(FLWF,0.5)/1E11} IS = 1.0E-16
* END CALC VALS
```

Voltage source:

```
* BEGIN PROG NSE NANO VOLT/RT-HZ
.SUBCKT VNSE 1 2
* BEGIN SETUP OF NOISE GEN - NANOVOLT/RT-HZ
* INPUT THREE VARIABLES
* SET UP VNSE 1/F
* NV/RHZ AT 1/F FREQ
.PARAM NLF = 60
* FREQ FOR 1/F VAL
.PARAM FLW = 0.1
* SET UP VNSE FB
* NV/RHZ FLATBAND
.PARAM NVR = 3.8
* END USER INPUT
* START CALC VALS
.PARAM GLF = {PWR(FLW,0.25)*NLF/1164}
.PARAM RNV = {1.184*PWR(NVR,2)}
.MODEL DVN D KF = {PWR(FLW,0.5)/1E11} IS = 1.0E-16
* END CALC VALS
```

Op-amp:

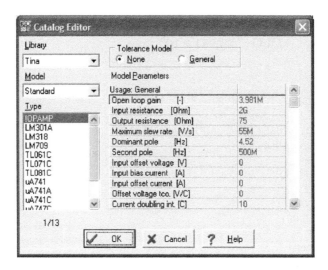

Simulation results match data sheet.

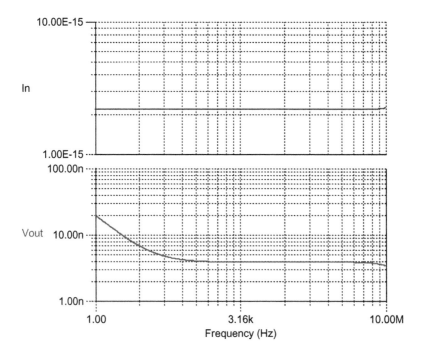

Chapter 5

5.1 $\quad V_{spect\_anal} = \sqrt{(10^{NdBm/10}) \cdot 1\,mW \cdot R} = \sqrt{(10^{-90/10}) \cdot 1\,mW \cdot 50\Omega} = 7.071\mu V$

$\quad V_{spect\_den} = \dfrac{V_{spect\_anal}}{\sqrt{K_n \cdot RBW}} = \dfrac{7.071\mu V}{\sqrt{1.128 \times 1\,kHz}} = 210.5\,nV/\sqrt{Hz}$

5.2 $\quad E_{noise} = \sqrt{E_{meas}^2 - E_{floor}^2} = \sqrt{(1.2\,mV)^2 - (0.2\,mV)^2} = 1.183\,mV$

Chapter 6

6.1 $\quad e_{nBB} = 3\,nV/\sqrt{Hz}$

$\quad e_{nBB\_WC} = e_{nBB} \cdot 1.1 = 3.3\,nV/\sqrt{Hz} \quad$ Worst case broadband noise

6.2 $\quad e_{n\_flicker} = 20\,nV/\sqrt{Hz} \qquad$ at $0.1\,Hz$

$\quad e_{n\_flicker\_WC} = 3 \cdot e_{n\_flicker} = 60\,nV/\sqrt{Hz} \quad$ at $0.1\,Hz$ Worst case flicker noise

6.3 $\quad i_{n\_opa827} = 2.2\,fA/\sqrt{Hz} \quad$ JFET input stage

$\quad i_{n\_opa227} = 400\,fA/\sqrt{Hz} \quad$ Bipolar input stage

6.4 Lower noise can be achieved by increasing the current supplied to the input differential stage. Typically noise is inversely proportional to the square root of current supplied to the input stage.

6.5 For most topologies, noise will typically increase proportionally to absolute temperature. This effect is normally not that noticeable because absolute temperature changes by a small amount relative to the operating range of a device.

Chapter 7

7.1 $R_{in} = 36k\dfrac{T}{i_n} = 36(1.381 \times 10^{-23}\, J/K)\dfrac{298.15\,K}{1(fA/\sqrt{Hz})} = 149\,G\Omega$

7.2

     a.   The large input impedance ($1\,M\Omega$) dominates the input noise $128\,nV/\sqrt{Hz}$. Any normal variation in the amplifiers' noise will be masked by the thermal noise of the resistor. Simulate this problem using the noise sources introduced in Chapter 4. Building a noise model will enable you to adjust noise levels in Part b of this problem. Simulation indicates noise at approximately $0.894\,mV$ RMS or $5.364\,mVpp$.

     b.   Adjust the noise sources so that $1/f$ noise is $3\times$ the typical value and broadband voltage noise is 10% greater than typical. Also adjust current noise so that it is $4\times$ the typical value. In this case, the $1\,M$ input resistor thermal noise dominates and so the output noise is relatively unaffected by changes in amplifier noise. Simulation indicates noise at approximately $0.897\,mV$ RMS or $5.382\,mVpp$.

     c.   The limit can be set at a level greater than the result computed in Part b. Popcorn noise normally is substantially larger than the maximum expected noise. One possible choice for a limit is $10\,mVpp$. This is roughly twice the expected maximum noise. Setting the limit too close to the expected noise can result in erroneously failing good devices.

     d.   Popcorn noise normally exhibits a dramatic change in base current of bipolar transistors. In this example the OPA827 has JFET inputs. Any popcorn noise will occur inside the device and will show up as voltage popcorn noise. Normally a large input impedance is useful in detecting popcorn noise because it amplifies the effect of the step on the input current. In this case however, the input impedance could actually mask the popcorn noise by adding large thermal noise. In this case eliminating the input impedance would make it easer to detect popcorn noise.

In this example, we looked at the absolute magnitude of the output noise to detect popcorn noise. Another approach is to look at the noise signal wave shape. Popcorn has a distinct wave shape. One way to do this is to look for peaks in the derivative of the noise signal.

Chapter 8

8.1 $f = \dfrac{1}{10\,days \cdot 24(h/day)60(min/h)60(s/min)} = 1.16\,\mu Hz$

8.2 $\text{Time} = \dfrac{1}{1\mu\text{Hz}} = 1 \times 10^6\,\text{s}$

$\text{Time\_days} = \dfrac{1 \times 10^6\,\text{s}}{60(\text{s/min})60(\text{min/h})24(\text{h/day})} = 11.6\,\text{days}$

8.3 $\text{BW}_{n333} = \dfrac{350\,\text{kHz} \cdot 1.57}{101} = 5.44\,\text{kHz}$

$E_{n333} = e_{nBB}\sqrt{\text{BW}_n} = (55\,\text{nV}/\sqrt{\text{Hz}}\sqrt{5.44\,\text{kHz}}) = 4.057\,\mu\text{V}$

8.4 $e_{fnorm} = 225\,\text{nV}/\sqrt{\text{Hz}}\sqrt{10\,\text{Hz}} = 712\,\text{nV}$

$\text{BW}_{n364} = \dfrac{7\,\text{MHz} \cdot 1.57}{101} = 109\,\text{kHz}$

$E_{nf} = 712\,\text{nV}\sqrt{\ln\left[\dfrac{109\,\text{kHz}}{(1.16\mu\text{Hz})}\right]} = 3.57\,\mu\text{V}$

$E_{nBB} = e_{nBB}\sqrt{\text{BW}_n} = 17\,\text{nV}/\sqrt{\text{Hz}}\sqrt{109\,\text{kHz}} = 5.613\,\mu\text{V}$

$E_{n364} = \sqrt{E_{nf}^2 + E_{nBB}^2} = \sqrt{(3.57\mu\text{V})^2 + (5.613\mu\text{V})^2} = 6.652\,\mu\text{V}$

Chapter 9

9.1 $\text{BW}_n = 35\,\text{kHz} \cdot 1.57 = 54.95\,\text{Hz}$

$E_{nBB} = 9\,\text{nV}/\sqrt{\text{Hz}}\sqrt{54.95\,\text{kHz}} = 2.11\,\mu\text{V}$

$e_{fnorm} = 25\,\text{nV}/\sqrt{\text{Hz}}\sqrt{1\,\text{Hz}} = 25\,\text{nV}$

$E_{nf} = 25\,\text{nV}\sqrt{\ln\left(\dfrac{\text{BW}_n}{1\,\text{Hz}}\right)} = 82.592\,\text{nV}$

$E_n = \sqrt{E_{nBB}^2 + E_{nf}^2} = \sqrt{(2.11\mu\text{V})^2 + (82.592\,\text{nV})^2} = 2.11\text{-}\mu\text{V RMS}$

9.2 A direct comparison of the voltage noise spectral density curves shows the two noise curves are about equal at 1 kHz. Below 1 kHz, the OPA331 has larger input noise, above 1 kHz, the OPA331 has lower noise. Note this analysis only looks at the input stage of the INA333. In low gains the output stage noise is also a factor.

9.3 $\quad f_H := \dfrac{1}{2\mu(1\mu F)(1k\Omega)} = 159.155 \, Hz \quad f_L : 1 \, Hz \quad R_{brdg} := 10 \, k\Omega$

$BW_n := f_H \cdot 1.57 = 249.873 \times 10^0 \, Hz \quad k_n := 1.381 \times 10^{-23} \, J/K \quad I_{at\_f} := 6\dfrac{pA}{\sqrt{Hz}}$

$i_{nBB} := 0.3 \, pA/\sqrt{Hz} \quad Noise\_Gain := 50 \quad T_n : 298.15 \, K \quad E_{fnorm} := 30 \, nV$

$I_{nBB} := i_{nBB} \cdot \sqrt{BW_n} = 4.742 \times 10^{-12} \, A$

$I_{fnorm} := I_{at\_f} \sqrt{f_L} = 6 \times 10^{-12} \, A$

$I_{nf} := I_{fnorm} \sqrt{\ln\left(\dfrac{BW_n}{f_L}\right)} = 14.098 \times 10^{-12} \, A$

$I_n := \sqrt{I_{nf}^2 + I_{nBB}^2} = 14.874 \times 10^{-12} \, A$

$E_{n\_rb} := \sqrt{4k_n(T_n)\left(\dfrac{R_{brdg}}{2}\right)(BW_n)} = 143.446 \times 10^{-9} \, V$

$E_{n\_i} := \sqrt{2\left(I_n \dfrac{R_{brdg}}{2}\right)^2 + 2(E_{n\_rb})^2} = 228.508 \times 10^{-9} \, V$

$E_{n\_opaBB} := 10 \, nV/\sqrt{Hz} \cdot \sqrt{BW_n} = 158.074 \times 10^{-9} \, V$

$E_{n\_opaf} := E_{fnorm}\sqrt{\ln\left(\dfrac{BW_n}{f_L}\right)} = 70.49 \times 10^{-9} \, V$

$E_{n\_opa} := \sqrt{E_{n\_opaBB}^2 + E_{n\_opaf}^2} = 173.079 \times 10^{-9} \, V$

$E_{n\_rti} := \sqrt{E_{n\_opa}^2 + E_{n\_i}^2} = 286.657 \times 10^{-9} \, V$

$E_{n\_out} := Noise\_Gain \cdot E_{n\_rti} = 14.333 \times 10^{-6} \, V$

Chapter 10

10.1 Photodiode capacitance decreases with increasing reverse bias voltage (see equation in Figure 10.4).

10.2
    a. Reduces noise peaking.
    b. Improves stability
    c. Does not affect signal bandwidth. Signal bandwidth is set by R_f and C_f.

10.3 Increase C_f. See Figure 10.18.

10.4

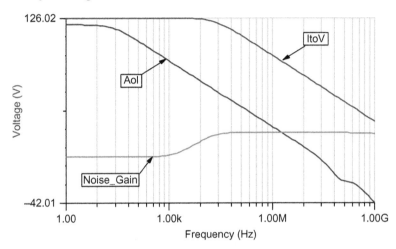

Chapter 11

11.1 $\quad k_b = 1.38 \times 10^{-23} \, \text{J/K}$

$\quad T_n = 298 \, \text{K}$

$\quad R_{sh} = 250 \, \text{M}\Omega$

$\quad i_j = \sqrt{\dfrac{4 k_b T_n}{R_{sh}}} = 8.112 \, \text{fA}/\sqrt{\text{Hz}}$

$\quad q = 1.602 \times 10^{-19} \, \text{C}$

$\quad I_D = 10 \times 10^{-9} \, \text{A}$

$\quad i_{sL} = \sqrt{2 q I_D}$

$\quad i_{n\_diode} \sqrt{i_j^2 + i_{sL}^2} = 57.182 \, \text{fA}/\sqrt{\text{Hz}}$

11.2

a. $R_f = 200\,k\Omega$ $C_f = 4\,pF$

$C_j = 40\,pF$ $C_{opa} = 9\,pF$

$C_i = C_j + C_{opa} = 49\,pF$

$f_c = 20\,MHz$

$f_p = \dfrac{1}{2\pi R_f C_f} = 199\,kHz$

$f_z = \dfrac{1}{2\pi R_f C_f} = 199\,kHz$

$f_i = \dfrac{C_f}{C_i + C_f} \cdot f_c = 1.51\,MHz$

$e_{nif} = 7\dfrac{nV}{\sqrt{Hz}}$

$e_{at\_f} = 55\,nV/\sqrt{Hz}$ $f_L = 10\,Hz$

$e_{fnorm} = e_{at\_f} \cdot \sqrt{f_L} = 174\,nV$

$f_f = \dfrac{e_{fnorm}^2}{e_{nif}^2} = 617\,Hz$

11.3

b. $E_{noe1} := \sqrt{e_{nif}^2 \cdot f_f \cdot \left(\dfrac{f_f}{f_L}\right)} = 1.367 \times 10^{-6}\,V$

$E_{noe2} := \sqrt{e_{nif}^2 \cdot (f_z - f_f)} = 839.92 \times 10^{-9}\,V$

$E_{noe3} := \sqrt{\left(\dfrac{e_{nif}}{f_z}\right)^2 \cdot \dfrac{f_p^3 - f_z^3}{3}} = 23.879 \times 10^{-6}\,V$

$E_{noe4} := \sqrt{\left(e_{nif} \cdot \dfrac{C_i + C_f}{C_f}\right)^2 \cdot (f_i - f_p)} = 106.177 \times 10^{-6}\,V$

$E_{noe5} := \sqrt{\dfrac{(e_{nif} \cdot f_c)^2}{f_i}} = 113.952 \times 10^{-6}\,V$

$E_{noe} := \sqrt{E_{noe1}^2 + E_{noe2}^2 + E_{noe3}^2 + E_{noe4}^2 + E_{noe5}^2} = 157.58 \times 10^{-6}\,V$

c. $i_{n\_opa} = 0.6\,\text{fA}/\sqrt{\text{Hz}}$

$i_{n\_diode} = 57.18\,\text{fA}/\sqrt{\text{Hz}}$

$i_{n\_total} = \sqrt{i_{n\_opa}^2 + i_{n\_diode}^2} = 57.18\,\text{fA}/\sqrt{\text{Hz}}$

$K_n = 1.57$

$BW_n = K_n f_p = 312\,\text{kHz}$

$E_{noi} = i_{n\_total} R_f \sqrt{BW_n} = 6.392\text{-}\mu\text{V RMS}$

d. $E_{noR} = \sqrt{4kTR_f BW_n} = 32.076\,\mu\text{V}$

$E_{no} = \sqrt{E_{noR}^2 + E_{noi}^2 + E_{noe}^2} = \sqrt{(32.076\mu\text{V})^2 + (6.392\mu\text{V})^2 + (157.58\mu\text{V})^2}$
$= 161\text{-}\mu\text{V RMS}$

11.4

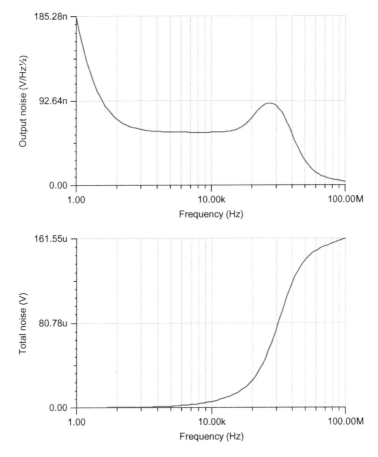

The thermal noise from the feedback resistor obscures the noise peak.

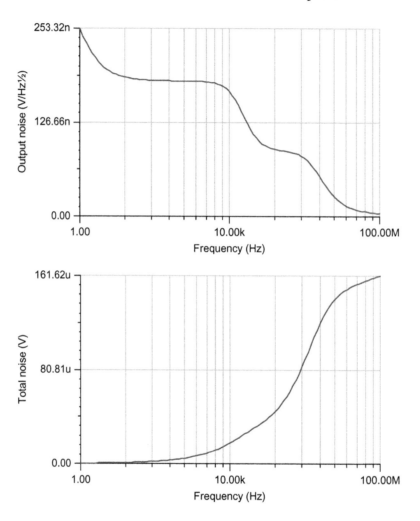

Index

Printed and bound by CPI Group (UK) Ltd, Croydon, CR0 4YY

03/10/2024

01040333-0001